DATE DUE

MAR 2 0			
GAYLORD			PRINTED IN U.S.A.

Solving Problems in Structures

Volume 2

Other titles in the Series

Solving Problems in Structures
Volume 2

P. C. L. Croxton BSc, PhD, MICE, CEng

L. H. Martin BSc, PhD, FICE, CEng

Longman
Scientific &
Technical

Copublished in the United States with
John Wiley & Sons, Inc., New York

Longman Scientific & Technical,
Longman Group UK Limited,
Longman House, Burnt Mill, Harlow,
Essex CM20 2JE, England
and Associated Companies throughout the world.

Copublished in the United States with
John Wiley & Sons, Inc., 605 Third Avenue,
New York, NY 10158

First published 1990

British Library Cataloguing in Publication Data
Croxton, P.C.L. (Peter Charles Lester), *1924–*
  Solving problems in structures.
  Vol. 2
  1. Structures. Analysis
  I Title   II. Martin, L.H. (Laurence Harold), *1928–*
  III Series
  624.1'71

ISBN 0-582-02355-6

Library of Congress Cataloging-in-Publication Data
(Available)

Set in 10/12 Compugraphic Times

Produced by Longman Group (FE) Limited
Printed in Hong Kong

Contents

Preface

This book deals with the solution of problems in Structural Analysis at an intermediate and advanced level and is intended primarily for students in the second and final years of a course leading to a degree in civil engineering at a university or polytechnic. Fully worked solutions are given to approximately 100 problems and outine solutions to approximately 80 problems intended for self assessment.

The relevant theory is also covered, but as the book is essentially concerned with problem solving formal proofs of theorems have been omitted. Otherwise the theoretical content is sufficient for the book to be used as a stand-alone text. An ideal companion text is Volume 1 of the same title which deals with elementary problems in statics and statically determinate structures.

Students of Structural Design and professional design engineers should find the computer programs at the end of the book of practical use. These programs are written in BASIC and can be run on almost any personal computer with a free memory of about 30k bytes. The methods of analysis used in the programs are fully described and examples of typical runs are given. The final program, which is based on the stiffness method of analysis, employs a compact storage technique enabling it to deal with structures of a practical size, and instructions are given for its expansion where memory capacity is available.

Exponential notation

Exponential notation is used throughout this book. For example

1234000, i.e. 1.234×10^6 is written as 1.234E6
0.001234, i.e. 1.234×10^{-3} is written as 1.234E-3.

This form, using E, is the form in which exponential notation would be written in a computer program. On an electronic calculator, E denotes pressing the EXP key (EEX on some calculators) by means of which numbers in this form can be entered directly.

Since the prefixes to SI units are in steps of 10^3, exponents which are multiples of 3, are preferred. For example, 123.4E3 or 0.1234E6 are preferable to 1.234E5.

1

Introduction

Methods of analysis

Volume 1 of *Solving Problems in Structures* dealt with statics and statically determinate structures. Volume 2 is chiefly concerned with the determination of elastic displacements and the solution of problems involving hyperstatic structures. Chapter 9 deals with the determination of collapse loads.

The elastic analysis of a hyperstatic structure can be approached in two basically different ways which may be summarised briefly as follows. In the flexibility approach the redundant restraints are the unknowns in the problem. The structure is made statically determinate by removing the restraints, which are then treated as external loads and determined from equations of compatibility at the restrained points. In the stiffness approach the external displacements of the joints are the unknowns. These are determined from equations derived by considering the equilibrium at the joints between the external loads and the internal forces and moments in the members.

Since the final equations have to be solved simultaneously the choice of a hand method of analysis will usually depend upon the number of unknowns in the problem. Consider for example the structures in Fig. 1.1.

Structure (a) is a pin-jointed frame with a single redundancy, requiring the solution of a single equation if a flexibility method is used. Now consider the stiffness approach. Since each joint has freedom to move in the x and y directions, except when restrained by a support, the total number of unknown joint displacements is 13. The remaining structures are rigidly jointed frames in which each free joint has three degrees of freedom, namely in the x, y and θ directions. However, in this type of structure, where flexure is predominant, the number of unknown joint displacements can usually be reduced for hand methods of analysis by ignoring axial deformations in the members. Structure (b) has one redundancy but eight joint displacements, reducing to five if axial deformations are ignored. Making the same simplification there are three redundancies and three joint displacements in structure (c), and nine redundancies but only five joint displacements in structure (d). For analysis by hand therefore, a flexibility method would be preferred for structures (a) and (b), a stiffness method for (d), while either method would be suitable for (c).

Both the flexibility and the stiffness approaches can be generalised by the use of matrix algebra. In particular the stiffness approach can readily be adopted for analysis by computer and has the additional advantage that the displacements at all the joints, as well as the member forces, are automatically output. In

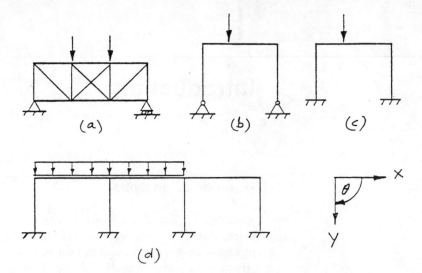

Figure 1.1

recent years many structural analysis programs have been developed for personal computers; and these are being increasingly used in design offices. The listing in BASIC of such a program is given at the end of Chapter 10. The need for some of the older methods which were developed to enable difficult or complicated problems to be solved by hand has thus been eliminated. Today it is unlikely that hand solution of problems involving more than three simultaneous equations would be considered practical, but for relatively trivial problems the preparation, checking, and entering into the computer of data can often be more time consuming than solving the problems by hand. Partial solution by a flexibility method in which the equations are developed by hand and solved by computer is a possible alternative to a complete solution by computer and a suitable program is given in Chapter 10. Aside from the practical considerations of the design office, however, the solution of problems by hand is essential if a thorough understanding of structural principles is to be acquired.

Of the hand methods, probably the most useful of those based on flexibility is the *Unit Load* method described in Chapter 7. It can be applied with almost equal facility to pin-jointed or rigidly jointed structures. Stiffness methods are not suitable for pin-jointed structures. For rigidly jointed structures the most useful stiffness method is the *Hardy Cross moment distribution* technique described in Chapter 5. It is particularly suited to the analysis of continuous beams and single storey portal frames, and with complicated beam loadings it is often better than the *Unit Load* method, provided that sway is not involved. The *Slope-Deflection* technique of Chapter 4 is a stiffness method which is particularly useful when the effect of an applied displacement is to be determined. The use of indirect model analysis, as described in Chapter 8, can sometimes provide a quick and easy practical solution to problems that present particular difficulties, such as, for example, structures with members that are curved or of continuously varying section.

For the determination of deflections the *Macaulay* method of Chapter 2 is probably the best for beams with relatively complicated loading, but a much quicker solution can be obtained by the *Area-Moment* method of Chapter 3 for simple loading cases. For other structures, with either pinned or rigid joints, the *Unit Load* method of Chapter 6 is recommended.

The principle of superposition

The principle of superposition enables a complicated problem to be subdivided into a number of simpler problems which can be solved separately and then re-combined. It can be stated generally as follows, *The combined effect of simultaneous actions on a structure is equal to the sum of the effects of each action taken separately.* In this context *effects* include displacements, member forces, and residual stresses; *actions* include external loads, applied displacements such as sinking supports, changes in temperature, and self straining actions such as welding. Some examples of how the principle can be applied are given in Fig. 1.2.

The conditions for full generality are that:

(1) throughout the application of the actions, whether separately or

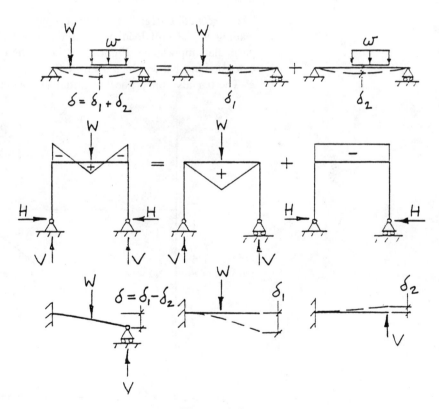

Figure 1.2

simultaneously, the relationship between external loads, displacements, and member forces remains linear and elastic

(2) the deformations are not gross.

Notable cases where the principle does not apply are in the analysis of slender compression members and in the determination of work or energy. In the first case lateral displacements are not linearly related to the axial loads. The second case may be demonstrated by considering the application of a new external load to a structure which is already subjected to other loads. The new load will do work by causing a displacement at its point of application, but as it will also produce displacements at the other load points the existing loads will also do work. Consequently the work done by the simultaneous application of loads is generally greater than the sum of the work done by each separate load. The exception is when none of the displacements produced by any load has a component in the direction of any other load. For example, assuming the simple theory of bending, the internal energies resulting from the stress resultants axial force, shear force, and bending moment in a beam do not interact, so the total energy may be obtained by simple summation.

Strain and complementary energy

The concept of energy has a number of applications in structural analysis, those leading to the unit load methods described in Chapters 6 and 7 being of particular importance. Figure 1.3 shows the relationship between force and extension in a bar to which a tensile force P is applied, producing an extension e. The bar also undergoes an initial extension λ due to thermal expansion.

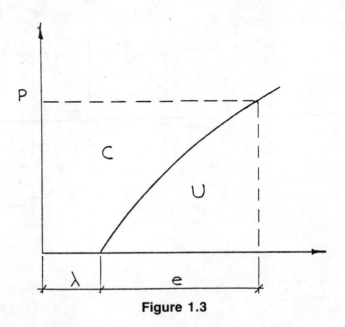

Figure 1.3

Strain energy is defined as

$$V = \int_0^e P \, de \qquad (1.1)$$

This is the area under the curve and is the work done by the force P stretching the bar. If the bar is elastic it is stored as potential energy and is recoverable, the bar acting as a stiff spring. The area above the curve, which includes the effect of the initial extension, is

$$C = P\lambda + \int_0^P e \, dP \qquad (1.2)$$

This is the *complementary energy*. Unlike strain energy it has no physical significance. For a constant force the integration can be performed by expressing e as PL/EA, giving

$$C = P\lambda + P^2L/2EA \qquad (1.3)$$

Under general loading the total strain or complementary energy in an element of the bar is given by the sum of the energies due to all the stress resultants acting at the time (see Principles of Superposition). For a linear elastic element of length dz, without initial extension, the strain and complementary energies are equal and are given by

Axial force:	$P^2dz/2EA$
Bending moment:	$M^2dz/2EI$
Shear force:	$S^2dz/2GA$
Torque:	$T^2dz/2GJ$

$$(1.4)$$

All the above expressions have the same form, namely *half the square of the stress resultant multiplied by the corresponding flexibility of the element*. Their derivation may be obtained from any standard text on Strength of Materials.

In a linear elastic structure the total strain energy stored in the members is equal to the work done by the external loads in deforming the structure, but this relationship can only be used in very simple cases for the determination of deflections. A much more general and useful result (see Chapter 6) is derived by equating the complementary work and energy.

Virtual work

Virtual work is a purely hypothetical concept which, although one of the most important in structural mechanics, tends to be neglected today as a primary method of analysis and is used mainly for derivation and proof. Consider a structure having external loads W_i with vectorially equivalent displacements δ_i at the joints, and internal forces P_j with vectorially equivalent displacements e_j in the members. Then, provided that the loads W_i are in equilibrium with the forces P_j and the displacements δ_i are compatible with the displacements e_j, the principle of virtual work states that

$$\underset{\text{joints}}{\sum W_i \delta_i} = \underset{\text{members}}{\sum P_j e_j} \qquad (1.5)$$

If the conditions of equilibrium and compatibility are satisfied neither of the sets of forces or displacements need actually exist, but may be only *virtual*. Combinations of virtual forces with real displacements and vice versa lead to indirect methods of analysis in which the final equations are transformed into a more easily managed form. For example in the matrix flexibility or *force* method (Chapter 10) the problem of forming compatibility equations is transformed into a simpler equilibrium problem by combining virtual forces with real displacements. Conversely a combination of virtual displacements with real forces leads to a stiffness or *displacement* method in which an equilibrium problem is transformed into one of compatibility. Examples are the use of virtual displacements in the construction of influence lines (Chapter 8) and the virtual work method of determining the collapse load of mechanisms (Chapter 9).

Worked examples

1.1 Complementary energy – non-linear material

Determine the complementary energy for a bar under uniform tension if the stress σ and the strain ϵ are related by $\epsilon = \sigma/E + \sigma^3/k$, where E and k are constants.

Solution

$$C = \int_0^P e \; dP$$

where

$$P = \sigma A \text{ and hence } dP = A \; d\sigma$$
$$e = \epsilon L = L(\sigma/E + \sigma^3/k)$$

Hence

$$C = AL \int_0^\sigma (\sigma/E + \sigma^3/k)d\sigma = AL(\sigma^2/2E + \sigma^4/4k)$$

1.2 Variable force

A uniform steel bar 30 m long is suspended vertically from one end. Young's modulus is 200E3 N/mm^2, the density of steel is 7850 kg/m^3, and the cross sectional area is 20 mm^2. Calculate the strain energy in the bar
  (a) when the bar hangs under its own weight
  (b) when an additional weight of 50 N is attached to the lower end.
  Show by equating real work and strain energy that in (a) the extension is that produced by a constant force equal to the average force in the bar, in (b) the total strain energy is the sum of the strain energies due to
    (i) the self-weight of the bar
    (ii) the 60 N force acting alone
    (iii) the work done by the average force in the bar, due to its self-weight, acting through the extension produced by the 50 N force.

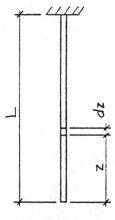

Figure 1.4

Solution (a) From Fig. 1.4 the force on an element of the bar is wz, where w is the weight per unit length, i.e.

$$w = 7850\text{E-}9 \times 20 \times 9.81 = 1.54\text{E-}3 \text{ N/mm}$$

Strain energy in element

$$\mathrm{d}U = P^2\mathrm{d}z/2EA = w^2z^2\mathrm{d}z/2EA$$

Total strain energy

$$U = w^2 \int_0^L z^2\mathrm{d}z/2EA = w^2L^3/6EA$$
$$= (1.54\text{E-}3)^2 \times (30\text{E}3)^2/(6 \times 200\text{E}3 \times 20)$$
$$= 2.67 \text{ Nmm } (2.67\text{E-}3 \text{ J})$$

Extension of element

$$\mathrm{d}e = P \,\mathrm{d}z/EA = wz \,\mathrm{d}z/EA$$

Total extension

$$e = w\int_0^L z \,\mathrm{d}z/EA = wL^2/2EA = wL/2 \times L/EA$$

This is the extension due to the average force ($wL/2$) in the bar.

Solution (b) Force on element $= (wz+50) \; N$

Hence

$$\mathrm{d}U = (wz+50)^2\mathrm{d}z/2EA$$
and $$U = \int_0^L (w^2z^2 + 50^2 + 100wz)\mathrm{d}z/2EA$$

On integrating, this can be written in the form $U = U_1+U_2+U_3$, where

$$U_1 = w^2L^3/6EA \text{ the strain energy due to the self-weight}$$

$U_2 = 50^2L/2EA$ the strain energy due to the 50 N force

$U_3 = 100wL^2/4EA = wL/2 \times 50L/EA$ the work done by the average self-weight force acting through the extension due to the 50 N force

Problems

1   Determine the strain energy stored in a bar 300 mm long, with cross sectional area 100 mm^2, when it is stretched 0.2 mm. Young's modulus is 210E3 N/mm^2.

Solution

$$U = Pe/2 = e^2EA/2L = 1.4\text{E}3 \text{ Nmm } (1.4 \text{ J})$$

2   Determine the strain and complementary energy in the bar in Problem 1 if the extension is due partly to a tensile force and partly to a rise in temperature of 30°C. The coefficient of linear expansion is 12E-6/°C.

Solution

$$C = P\lambda + P^2L/2EA, \quad \lambda = 30 \times 12\text{E-}6 \times 300 = 0.108 \text{ mm}$$
$$P = eEA/L = (0.2 - 0.108) \times 210\text{E}3 \times 100/300 = 6.44\text{E}3 \text{ N}$$
Hence $C = 992$ Nmm (0.992 J)

3   In a tensile test the following corresponding forces and extensions were recorded.

P (kN)	0	7.5	15
e (mm)	0	1.5	4

(a)   Assuming the relationship $e = aP^2 + bP + c$ determine the complementary energy when $P = 15$ kN.

(b)   What is the corresponding strain energy?

Solution

$$a = 8.889\text{E-}3, \quad b = 133.3\text{E-}3, \quad c = 0$$
$$C = \int_0^{15} (aP^2 + bP)\mathrm{d}P = 25 \text{ Nm } (J)$$
$$U = Pe - C = 35 \text{ Nm } (J)$$

4   By equating strain energy and real work show that the maximum deflection of a light cantilever of length L carrying a point load W at the free end is $WL^3/3EI$.

Solution

$$U = \int W^2z^2 \, \mathrm{d}z/2EI = W^2L^3/6EI = w\delta/2. \quad \text{Hence } \delta = WL^3/3EI.$$

2

Deflection of beams: Macaulay's method

Introduction

It was shown in Volume 1 of *Solving Problems in Structures* that the following relationship exists in a beam:

$$1/R = d\theta/ds = M/EI \tag{2.1}$$

where R is the radius of curvature, M the bending moment, and EI the flexural rigidity. From Fig. 2.1

$$\theta = -dy/dz \tag{2.2}$$

and, since $d\theta/ds \approx d\theta/dz$ when the curvature is small,

$$d^2y/dz^2 = -M/EI \tag{2.3}$$

This is the differential equation of bending. Successive integrations give the equations of the slope and deflection along the beam, provided that M/EI can be expressed as an integrable function of z. Discontinuities affecting the second moment of area or Young's modulus can be handled by integrating in stages. Difficulties involving discontinuities in the loading can be overcome by Macaulay's method in which the equation for the bending moment is formed at a point on the beam beyond which no discontinuities occur.

For example, in Fig. 2.2 at point Z on the beam

$$M = R_Az - W_1[z-a] - W_2[z-b] - w[z-c]^2/2 \tag{2.4}$$

The terms in square brackets are *step functions* which take the value of zero when they become negative. On integration they are not expanded, but are treated as single variables. The resulting errors are constants which are automatically corrected when the arbitrary constants of integration are determined from the boundary conditions. Combining equations (2.3) and (2.4) and using the notation y' and y'' for the differential coefficients dy/dz and d^2y/dz^2.

$$EIy'' = -M = -R_Az + W_1[z-a] + W_2[z-b] + w[z-c]^2/2 \tag{2.5}$$

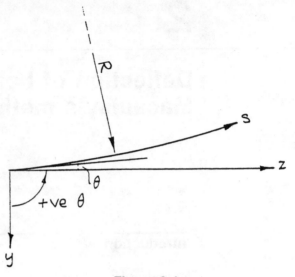

Figure 2.1

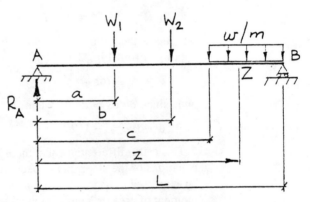

Figure 2.2

Integrating twice

$$EIy' = -R_A z^2/2 + W_1[z-a]^2/2 + W_2[z-b]^2/2 + \\ w[z-c]^3/6 + A \tag{2.6}$$

and $EIy = -R_A z^3/6 + W_1[z-a]^3/6 + W_2[z-b]^3/6$
$$+ w[z-c]^4/24 + Az + B \tag{2.7}$$

The boundary conditions are that the deflection is zero at the ends of the beam. The arbitrary constants A and B can therefore be determined by substituting $y = 0$ when $z = 0$ and L in equation (2.7). Note that when $z = 0$ all the step functions become negative and therefore vanish; when $z = L$ they are all positive and remain in the expression.

The method is straightforward but tedious. However, it is probably the best method for statically determinate beams when the loading is complicated, or maximum deflections are required, or a general expression for the deflected shape is needed.

Worked examples

2.1 Point loads

A simply supported beam AB of uniform section has a span of 10 m and carries point loads of 2 kN and 4 kN at distances of 4 m and 8 m from A respectively.

(a) Derive an expression for the deflected shape of the beam.

(b) If the second moment of area is 15E6 mm^4 and Young's modulus is 200 kN/mm^2, find the central deflection as a proportion of the span.

(c) What would be the percentage error if this deflection were taken to be the maximum?

Solution − *Fig. 2.3* (a) (1) Find the reaction at A:

$$R_A = (2 \times 6 + 4 \times 2)/10 = 2 \text{ kN}$$

(2) Form the differential equation of bending at a point beyond the 4 kN load

$$EIy'' = -M = -2z + 2[z-4] + 4[z-8]$$

(3) Integrate twice

$$EIy' = -z^2 + [z-4]^2 + 2[z-8]^2 + A$$
$$EIy = -z^3/3 + [z-4]^3/3 + 2[z-8]^3/3 + Az + B$$

(4) Determine A and B using the boundary conditions: $y = 0$ when $z = 0$ and 10.

Putting $z = 0$ all the step functions vanish, so $B = 0$
Putting $z = 10$

$$0 = -1000/3 + 216/3 + 16/3 + 10A, \text{ from which } A = 25.6$$

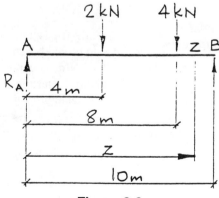

Figure 2.3

Hence

$$EIy = -z^3/3 + [z-4]^3/3 + 2[z-8]^3/3 + 25.6z$$

Solution (b) Putting $z = 5$ for the central deflection the last step function vanishes and

$$EIy = -125/3 + 1/3 + 25.6 \times 5 = 86.67$$

Care must now be taken with units. In the above equation the right hand side is in kNm^3, EI must therefore be calculated in kNm^2 to be consistent and give the deflection in m.

Hence

$$EI = 200 \times 15E6 = 3E9 \ kNmm^2 = 3E3 \ kNm^2$$

and $y = 86.67/3E3 = 0.02889$ m, i.e. $1/346$ of the span

Solution (c) Except when it is at the free end of a cantilever, the maximum deflection occurs at a turning point in the deflected shape, i.e. when $EIy' = 0$.

Hence

$$-z^2 + [z-4]^2 + 2[z-8]^2 + 25.6 = 0$$

As there are step functions in the expression the position of the maximum must be estimated. In the case of a single point load on a simply supported beam it is always between the load and mid-span; so that in this case it must be between the two loads. The second step function therefore vanishes, giving

$$-z^2 + (z-4)^2 + 25.6 = 0, \text{ from which } z = 5.2,$$

Substituting for z into the deflection equation $y = 0.02894$. The error incurred by taking the central deflection as the maximum is therefore 0.18%.

In design the central deflection of a simply supported beam subjected to uniformly distributed or point loads is usually taken as the maximum, the error always being small.

2.2 Distributed load continuous to support

> A beam ABC is 12 m long and is simply supported at A and C. A point load of 24 kN acts at B, 4 m from A, and a uniformly distributed load of 6 kN/m is continuous from B to C. If the flexural rigidity is 50E3 kNm^2, determine the slope and deflection at mid-span. State where, in relation to mid-span, the maximum deflection occurs.

Solution — Fig. 2.4 (1) Reaction $R_A = (24 \times 8 + 6 \times 8 \times 4)/12 = 32$ kN
(2) Form the differential equation of bending with z in BC.

i.e. $EIy'' = -M = -32z + 24[z-4] + 6[z-4]^2/2$

Integrating:

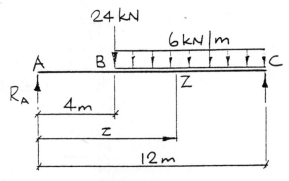

Figure 2.4

$$EIy' = -16z^2 + 12[z-4]^2 + [z-4]^3 + A$$
$$EIy = -16z^3/3 + 4[z-4]^3 + [z-4]^4/4 + Az + B$$

(3) The boundary conditions are $y = 0$ when $z = 0$ and 12.

Hence

$$B = 0$$

and $A = \{16 \times 12^3/3 - 4 \times 8^3 - 8^4/4\}/12 = 512$

(4) At mid-span $z = 6$ m

Hence

$$y = \{-16 \times 6^3/3 + 4 \times 2^3 + 2^4/4 + 512 \times 6\}/50E3 = 0.0392 \text{ m}$$

and $y' = \{-16 \times 6^2 + 12 \times 2^2 + 2^3 + 512\}/50E3 = -160E\text{-}6 \text{ rad}$

(5) Since y' is negative the slope is upwards. It is also very small. Therefore the maximum deflection must occur just to the left of mid-span.

2.3 Discontinuous distributed load

Determine the deflections at mid-span and at the free end of the beam in Fig. 2.5 $EI = 50E3$ kNm^2.

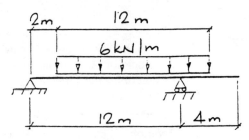

Figure 2.5

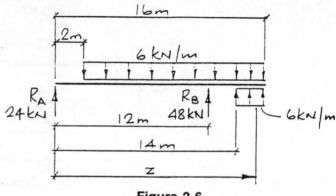

Figure 2.6

Solution — Fig. 2.6 (1) Find both reactions, using Fig. 2.5.

$$R_A = 6 \times 12 \times 4/12 = 24\,\text{kN}, \quad R_B = 72 - 24 = 48\,\text{kN}$$

(2) A valid bending moment expression using step functions cannot be written if the load is not continuous to the end of the beam. It must therefore be extended and the effect of this nullified by an upward load of the same intensity, as shown. The differential equation may then be formed. The reaction R_B is treated as an upwardly acting point load.

$$EIy'' = -M = -24z + 6[z-2]^2/2 - 48[z-12] - 6[z-14]^2/2$$
$$EIy' = -12z^2 + [z-2]^3 - 24[z-12]^2 - [z-14]^3 + A$$
$$EIy = -4z^3 + [z-2]^4/4 - 8[z-12]^3 - [z-14]^4/4 + Az + B$$

(3) The boundary conditions are $y = 0$ when $z = 0$ and 12, giving $B = 0$, $A = 367.7$.

(4) At mid-span $z = 6$. Substituting for EI and transposing,

$$y = \{-4 \times 6^3 + 4^4/4 + 367.7 \times 6\}/50E3 = 0.0281\,\text{m}$$

(5) At the free end $z = 16$.

$$y = \{-4 \times 16^3 + 14^4/4 - 8 \times 4^3 - 2^4/4$$
$$+ 367.7 \times 16\}/50E3 = -0.0283\,\text{m}$$

This problem is typical of the kind for which Macaulay's method provides the most convenient solution. The reader is recommended to try the alternative methods described later.

2.4 Beam loaded by a couple

A simply supported beam 10 m long is loaded by an anticlockwise couple of 30 kNm at mid-span. If the flexural rigidity is 20E3 kNm² determine the deflection at the load point and the maximum upward and downard deflections.

Solution — Fig. 2.7 (1) Reactions: $300/10 = 30\,\text{kN}$, acting as shown
(2) Since the deflected shape is skew-symmetrical, the deflection at mid-

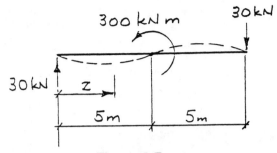

Figure 2.7

span must be zero, and the upward and downward deflections are equal. It is therefore only necessary to analyse half the beam.

(3) The differential equation of bending for the left hand side is

$$Ely'' = -30z$$

Integrating

$$Ely' = -15z^2 + A$$
$$Ely = -5z^3 + Az + B$$

The boundary conditions are $y = 0$ when $z = 0$ and 5, giving $B = 0$, $A = 125$.

(4) For the maximum downward deflection,

$$Ely' = -15z^2 + 125 = 0, \text{ from which } z = 2.887 \text{ m}$$

and $y = \{-5z^3 + 125z\}/20E3$

Substituting for z, $y_{max} = 0.012$ m.

2.5 Asymmetrical beam with couple

Repeat Example 2.4 with the couple 7 m from the left hand support.

Solution − Fig. 2.8 (1) Reaction: $R_A = 30$ kN

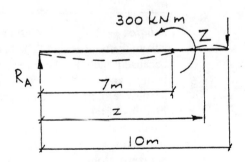

Figure 2.8

(2) Since the beam is no longer symmetrical, form the differential equation at point Z, i.e. to the right of the couple. The bending moment is given by

$$M = 30z - 300$$

However, in order to preserve the generality of the expression, a step function raised to the power of zero is used, thus

$$EIy'' = -M = -30z + 300[z-7]^0$$

Integrating,

$$EIy' = -15z + 300[z-7] + A$$
$$EIy = -5z^3 + 150[z-7]^2 + Az + B$$

The boundary conditions $y = 0$ when $z = 0$ and 10 give $B = 0$ and $A = 365$.
(3) At the load point $z = 7$.

Hence

$$y = \{-5\times7^3 + 365\times7\}/20\text{E3} = 0.042 \text{ m}$$

(4) For maximum downward deflection ($z < 7$), solve

$$EIy' = -15z^2 + 365 = 0$$

from which $z = 4.933$ m and the deflection is

$$y = \{-5z^3 + 365z\}/20\text{E3} = 0.060 \text{ m}$$

(5) For maximum upward deflection ($z > 7$), solve

$$EIy' = -15z^2 + 300(z-7) + 365 = 0$$

This equation has no real roots, so there is no turning point to the right of the load. Hence there is no upward deflection.

2.6 Propped cantilever

A propped cantilever ABC is built in at A and propped at C. The beam has a span of 9 m and carries a point load of 18 kN at B, 6 m from A. Given the second moment of area 2500 cm⁴ and Young's modulus 200 kN/mm², determine the force in the prop and the maximum deflection.

Solution — Fig. 2.9 With cantilevers it is generally preferable to measure z from the built-in end. The bending moment expression is more complicated because it includes the force and moment reactions at the built-in end, but as both y' and y are zero when z is zero, both arbitrary constants of integration are also zero. In the case of a propped cantilever the reactions must be expressed in terms of the propping force, which may then be determined directly from the third boundary condition that $y = 0$ at the prop. If z is measured from the free end it is necessary to solve a pair of simultaneous equations to find the non-zero arbitrary constant and the propping force.
(1) Reactions: $V = 18-P$, $M = 108-9P$

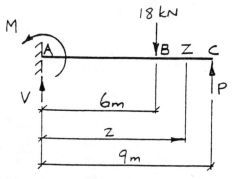

Figure 2.9

(2) $EIy'' = M - Vz + 18[z-6]$
$EIy' = Mz - Vz^2/2 + 9[z-6]^2 + A$
$EIy = Mz^2/2 - Vz^3/6 + 3[z-6]^3 + Az + B$

(3) Boundary conditions:

Putting $y' = y = 0$ when $z = 0$ yields $A = B = 0$
Putting $y = 0$ when $z = 9$ and substituting for M and V gives

$$0 = (108-9P) \times 9^2/2 - (18-P) \times 9^3/6 + 3 \times 3^3$$

from which $P = 9.333$ kN and hence $V = 8.667$ kN, $M = 24$ kNm.

(4) For maximum deflection $y' = 0$. Assuming that this occurs when $z < 6$,

$$0 = 24z - 4.333z^2, \text{ from which } z = 5,538 \text{ m}$$

Substitution into the deflection expression gives

$$EIy_{max} = 12 \times (5.538)^2 - 1.444 \times (5.538)^3 = 122.7 \text{ kNm}^3$$
$$= 122.7\text{E9 kNmm}^3$$
$$EI = 200 \times 2500\text{E4} = 5\text{E9 kNmm}^2$$

Hence

$$y_{max} = 122.7/5 = 24.5 \text{ mm}$$

2.7 Symmetrical beam

A simply supported beam has a span of 12 m and carries a uniformly distributed load of 12 kN/m over the central 6 m. If $EI = 60\text{E3}$ kNm^2 and is constant, calculate the maximum deflection.

Solution – Fig. 2.10 (1) Since the beam is truly symmetrical it is known that zero slope, and hence maximum deflection, occur at the centre. The differential equation of bending may therefore be formed for $z < 6$, thus

$$EIy'' = -36z + 12[z-3]$$
$$EIy' = -18z^2 + 6[z-3]^2 + A$$
$$EIy = -6z^3 + 2[z-3]^3 + Az + B$$

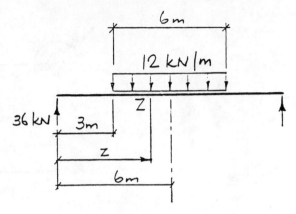

Figure 2.10

(2) Boundary conditions:

$y' = 0$ when $z = 6$ gives $A = 594$
$y = 0$ when $z = 0$ gives $B = 0$

(3) Putting $z = 6$ and substituting for EI gives $y_{max} = 0.0387$ m.

Superposition of standard cases

Table 2.1

Standard case	Deflection		Slope	
	1/48	centre	1/16	end
	$(3\alpha - 4\alpha^3)/48$	centre		
	5/384	centre	1/24	end
	1/3	end	1/2	end
	1/8	end	1/6	end
	1/2	end	1	end
	1/8	centre	1/2	end
Deflection = coefficient $\times$ WL^3/EI or ML^2/EI				
Slope = coefficient $\times$ WL^2/EI or ML/EI				

Rapid solutions to a number of problems can be obtained by superposition of standard cases. Table 2.1 gives some of the most common cases. Since the process can require repeated evaluation of a formula with different values for the variables, a calculator with the capability of storing formulae is especially useful.

Worked examples

2.8 Variable loading

Figure 2.11 shows the intensity in kN/m of the symmetrical load on a simply supported beam at intervals of 3 m. Given that EI is 2.83E6 kNm^2 obtain an approximate value for the central deflection.

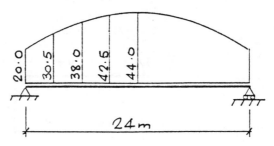

Figure 2.11

Solution (1) Represent the load approximately as eight point loads acting at the centre of each interval. The value of each load is given by

W = length of interval × mean of intensities on either side

For example the first load has the value of $3 \times (20.0+30.5)/2 = 75.75$ kN, acting at 1.5 m from the support.

(2) Tabulate values of load W in kN and distance a in m from the support, as follows.

No.	W	a	$\alpha = a/L$
1	75.8	1.5	0.0625
2	102.8	4.5	0.1875
3	120.8	7.5	0.3125
4	129.8	10.5	0.4375

(3) Evaluate $S = \Sigma W(3\alpha - 4\alpha^3)$ — from Table 2.1 — giving 294.6.

Then $\delta = 2S \times L^3/48EI = 2 \times 294.6 \times 24^3/(48 \times 2.83\text{E}6) = 59.96\text{E-}3$ m
Accurate result = 60.33E-3 m

2.9 Propped cantilever

A beam ABC is built in at A and simply supported at C, and carries a single point load at B. If the lengths of AB and BC are a and b respectively, and EI is constant, determine the reaction at C.

Solution — Fig. 2.12 The problem can be solved by the superposition of two load-cases, as shown, then, using the standard results from Table 2.1,

In case 1: $\delta_{C1} = \delta_B + b\theta_B = Wa^3/3EI + bWa^2/2EI$
In case 2: $\delta_{C2} = P(a+b)^3/3EI$

For compatibility $\delta_{C1} = \delta_{C2}$

Hence

$$P = Wa^2(2a+3b)/2(a+b)^3$$

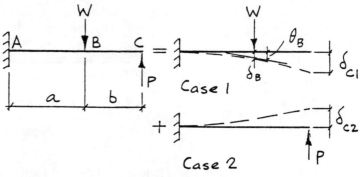

Figure 2.12

Problems

1 Prove the case in Table 2.1 for a simply supported beam with a point load off-centre.

Solution — refer to the figure in Table 2.1
$EIy'' = -W(1-\alpha)z + W[z-\alpha L]$
(Boundary conditions: $y = 0$ when $z = 0$ and $z = L$)
Hence $y = W\{-(1-\alpha)z^3+[z-\alpha L]^3+L^2z(2\alpha-3\alpha^2+\alpha^3)\}/6EI$
Putting $z = L/2$, $y = WL^3(3\alpha-4\alpha^3)/48EI$ (QED)

2 A cantilever is 6 m long and carries a uniformly distributed load of 30 kN/m for a distance of 4 m from the built-in end. Determine the deflection at the free end. EI is 80E3 kNm².

Solution — Fig. 2.13
This problem can either be solved by measuring z from the free end

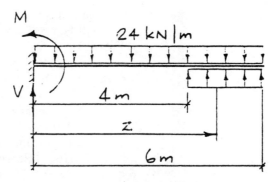

Figure 2.13

and forming the differential equation at a point in the load, or by starting from the fixed end and using dummy loads. In the first case neither arbitrary constant is zero; in the second both are zero. Try both methods. The solution by the second is given below.

$V = 24 \times 4 = 96\,\text{kN}, \quad M = 96 \times 2 = 192\,\text{kNm}$
$EIy'' = M - Vz + 12z^2 - 12[z-4]^2$
(Boundary conditions: $y' = y = 0$ when $z = 0$)
$EIy = 96z^2 - 16z^3 + z^4 - [z-4]^4$
Putting $z = 6$, $y = 1280/80E3 = 0.016\,\text{m}$

3 A beam ABCD of total length 16 m has a uniform section and rests on simple supports at A and C, 12 m apart. The beam carries an anticlockwise couple of 480 kNm at B, the mid-point of AC and a point load of 60 kN at the free end D. If Young's modulus is 200 kN/mm^2 and the second moment of area is 1.9E9 mm^4, determine the maximum upward and downward deflections.

Solution — Fig. 2.14

(1) The deflected shape will be similar to that shown by the dotted line, with the maximum upward deflection between B and C. The

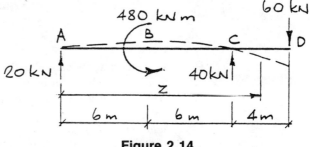

Figure 2.14

maximum downward deflection will almost certainly be at D, but it will be necessary to test between A and B.

$$(2) \quad EIy'' = -20z + 480[z-6]^0 - 40[z-12] \qquad (1)$$
$$EIy' = -10z^2 + 480[z-6] - 20[z-12]^2 + A \qquad (2)$$
$$EIy = -10z^3/3 + 240[z-6]^2 - 20[z-12]^3/3$$
$$+ Az + B \qquad (3)$$

Boundary conditions: $y = 0$ when $z = 0$ and 12. Hence $B = 0$ and $A = -240$

Setting equation (2) to zero for turning points:
$z < 6$ no real roots. Hence no downward deflection between A and B.
$6 < z < 12 \quad z = 7.752$ m
$EI = 200 \times 1.9E9 \times 1E-6 = 380E3$ kNm^2
Putting $z = 7.752$ and 16 in equation (3) gives maximum deflections:
0.007 m upwards, 0.016 m downwards.

4 A cantilever ABC is built in at A and propped at C, and carries a uniformly distributed load of intensity w covering the whole beam. If $AB = 2L$ and $BC = L$, determine the force in the prop.

Solution — Fig. 2.15
$$EIy'' = M - Vz + wz^2/2 - P[z-2L]$$
$$EIy' = Mz - Vz^2/2 + wz^3/6 - P[z-2L]^2/2 + A$$
$$EIy = Mz^2/2 - Vz^3/6 + wz^4/24 - P[z-2L]^3/6 + Az + B$$
By statics $V = 3WL - P$, $M = 9wL^2/2 - 2PL$
Boundary conditions: $y' = y = 0$ when $z = 0$, giving $A = B = 0$
$\qquad\qquad\qquad\qquad y = 0$ when $z = 2L$,
$\qquad\qquad\qquad\qquad$ *giving* $P = 17wL/8$

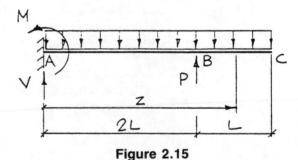

Figure 2.15

5 Determine the central deflection δ of a beam of length L and flexural rigidity EI, built in at both ends, and carrying a uniformly distributed load of intensity w and a central point load W.

Solution — Fig. 2.16
Consider as the sum of two simply supported beams loaded

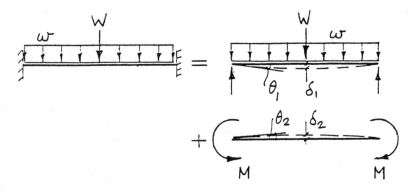

Figure 2.16

respectively with the applied loads and the fixed end moments, as shown. Examine the compatibility at the ends and at mid-span. Use Table 2.1.

$\theta_1 = \theta_2$: $WL^2/16EI + wL^4/24EI = ML/2EI$
            Hence $M = (3WL + 2WL^2)/24$

$\delta = \delta_1 - \delta_2$: Hence $\delta = WL^3/48EI + 5wL^4/384EI$
            $- ML^2/8EI = (2WL^3 - wL^4)/384EI$

6 The two beams in Fig. 2.17 were connected as shown by a light chain, before any loads were applied. Given the second moments of area indicated, determine:

(a) the maximum value of w if the chain is not to become slack
(b) the tension in the chain when $L = 4\,\text{m}$, $w = 16\,\text{kN/m}$, $W = 300\,\text{kN}$.

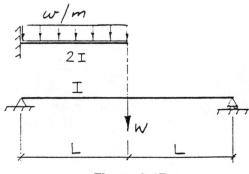

Figure 2.17

Solution — Table 2.1

(a) The chain is slack when $\delta_{\text{upper}} \geq \delta_{\text{lower}}$, i.e. when
$wL^4/(8E \times 2I) \geq W \times (2L)^3/48EI$
Hence $w_{\text{max}} = 8W/3L$

(b) Letting T = chain tension,
$\delta_{upper} = (wL^4/8 + TL^3/3)2EI$, $\delta_{lower} = (W-T) \times (2L)^3/48EI$
Substituting for L, w and W, and setting $\delta_{upper} = \delta_{lower}$,
$T = 138\,kN$

7 A uniform simply supported beam, propped at mid-span, carries a uniformly distributed load of 960 N/m over a span of 6 m. Young's modulus and second moment of area are 180 kN/mm^2 and 40E6 mm^4 respectively. Determine the load taken by the prop when
  (a) the beam is propped level with the supports
  (b) the prop settles 2 mm relative to the supports.

Solution — Fig. 2.18 and Table 2.1
  (a) $\delta_1 = \delta_2$, i.e. $5WL^3/384EI = PL^3/48EI$, where
      $W = 0.96 \times 6 = 5.76\,kN$
      Hence $P = 5W/8 = 3.6\,kN$
  (b) $\delta_1 = \delta_2 = 2\,mm$, i.e. $5WL^3/384EI = PL^3/48EI + 2$,
      from which $P = 0.4\,kN$

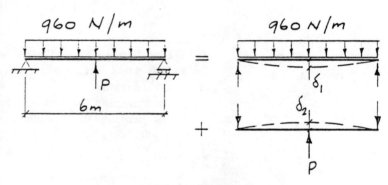

Figure 2.18

3

Area and area-moment methods of analysis

Introduction

Area and area-moment are hand calculation methods of structural analysis which produce solutions to certain simple types of static and hyperstatic elastic structures where members are predominantly in bending. It is particularly useful for analysing the behaviour of single members and producing general relationships which are of use in more advanced methods, e.g. fixed end moments, moment distribution factors, and slope deflection equations. Application of the method requires an understanding of the relationships between moments, slopes and deflections for a member, but it becomes laborious with more complicated hyperstatic structures.

The basis of the method is the differential equation for a member relating section properties and bending moment for a member in the elastic stage of behaviour (see equation 2.3), i.e.

$$\mathrm{d}^2y/\mathrm{d}z^2 = -M/EI \tag{3.1}$$

where M is the bending moment at a section distance z from the origin, E is Young's modulus, I is the second moment of area of the section about the neutral axis, y is the deflection perpendicular to the length of the members, and z is the distance along the member.

If equation (3.1) is integrated between limits z_1 and z_2 then in general terms the change in slope is

$$(\mathrm{d}y/\mathrm{d}z)_{z2} - (\mathrm{d}y/\mathrm{d}z)_{z1} = -\int_{z1}^{z2} M\ \mathrm{d}z/EI$$

For general application this can be written as

$$[\mathrm{d}y/\mathrm{d}z]_{z1}^{z2} = -[A/EI]_{z1}^{z2} \tag{3.2}$$

where A is the area of the bending moment diagram between limits z_1 and z_2.

If both sides of equation (3.1) are multiplied by z and integrated between limits z_1 and z_2 then the relationship between slope and deflection is

$$\int_{z1}^{z2} z(\mathrm{d}_2y/\mathrm{d}z^2)\ \mathrm{d}z = -\int_{z1}^{z2} Mz\ \mathrm{d}z/EI$$

For general application this can be written as

$$[z\ \mathrm{d}y/\mathrm{d}z - y]_{z1}^{z2} = -[A\bar{z}/EI]_{z1}^{z2} \tag{3.3}$$

where $A\bar{z}$ is the first moment of the area of the bending moment about the origin between limits z_1 and z_2.

As can be seen the method incorporates bending moment diagrams and it is necessary to be able to sketch these before attempting to use it. Greater understanding of the fundamentals of the method can be obtained by working through Macaulay's method (Chapter 2). A bending moment diagram for a member can be represented by standard shapes, and some common areas and centroids of area are shown in Table 3.1.

The sign convention for bending moment diagrams, slopes and deflections is based on the right hand screw rule as described in Volume 1 of *Solving Problems in Structures*.

Table 3.1

Shape	Area	Position of centroid
triangle	$\frac{1}{2} L H$	
half parabola	$\frac{2}{3} L H$	
compliment of half parabola	$\frac{1}{3} L H$	

Worked examples

3.1 Slope and deflection at the end of a cantilever

Calculate the slope and deflection at the end of the cantilever which supports a point load as shown in Fig. 3.1(a)

Solution This is a simple statically determinate structure. Draw the *M/EI* diagram for the cantilever as shown in Fig. 3.1(c). The shape of the bending

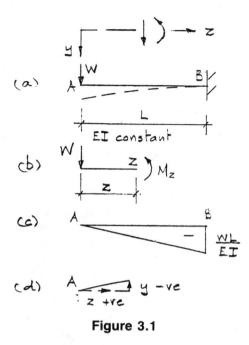

Figure 3.1

moment diagram is triangular and the negative sign is in accordance with the right hand screw rule. Positive directions of axial force, shear force and bending moment are shown diagramatically above the cantilever. At a section distance z from A (see Fig. 3.1(b))

$$Z\} + M_z + Wz = 0; \quad \text{hence } M_z = -Wz$$

To determine the slope at A, apply equation (3.2) between A and B with the original at A:

$$[dy/dz]_0^L = -[A/EI]_0^L$$

and $\quad [0 - \theta_A] = -[(1/2) \times L \times (-WL/EI)]$

hence

$$\theta_A = -WL^2/(2EI) \tag{1}$$

The slope at A is negative which means that the end of the cantilever is deflected as shown in Fig. 3.1(d). In Fig. 3.1(d) the slope is defined by z positive (to the right) and y negative (upwards), which results in a negative slope. This agrees with a common sense interpretation of the physical behaviour of the cantilever. In other problems the answer is not obvious and a strict adherence to the sign convention is necessary. Notice that one of the slopes (θ_B) is zero and the left hand side of the equation reduces to θ_A.

To determine the deflection at A, apply equation (3.3) between A and B with the origin at A:

$$[z \, dy/dz - y]_0^L = -[A\bar{z}/EI]_0^L$$

and

$$[L \times 0 - 0] - [0 \times \theta_A - y_A] = -[(1/2) \times L \times (-WL/EI) \times (2L/3)] \text{ (2)}$$

Hence, combining (1) and (2)

$$y_A = +WL^3/(3EI)$$

The positive sign means that the deflection is downwards, which is obviously correct. Notice that by choosing the origin at A the left hand side of the equation reduces to the deflection at A.

3.2 Fixed end moments

Calculate the fixed end moments and deflection at centre span for the fixed end beam, shown in Fig. 3.2(a), which supports a point load W at centre span.

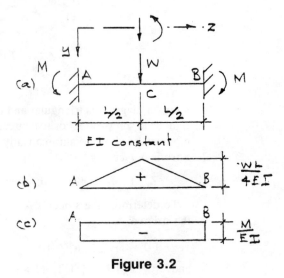

Figure 3.2

Solution Draw the M/EI diagrams shown in Figs 3.2(b) and (c). The signs of the bending moments are known, or assumed, and are in accordance with the right hand screw rule. For convenience bending moment diagrams are reduced to simple cases. Figure 3.2(b) is the diagram for a point load on a simply supported span. Figure 3.2(c) is the diagram for equal end moments M applied to a simply supported span.

To determine the fixed end moments, apply equation (3.2) between A and B with the origin at A:

$$[dy/dz]_0^L = -[A/EI]_0^L$$

and

$$[0 - 0] = -\{[(1/2) \times L \times (+WL/4EI)] + [L \times (-M/EI)]\}$$

hence

$$M = +WL/8$$

This result is useful in the methods of moment distribution and slope deflection (see Table 4.1). The bending moment is positive because the assumed directions of the end moments are correct. Notice that the second term on the right hand side of the equation is added when dealing with bending moment diagrams and also notice that the left hand side of the equation reduces to zero which enables M to be determined directly.

To determine the deflection at mid-span, apply equation (3.3) between A and centre span (C) with the origin at A,

$$[z \, dy/dz - y]_0^{L/2} = -[A\bar{z}/EI)_0^{L/2}$$

and

$$[(L/2) \times 0 - y_C] - [0 \times 0 - 0]$$
$$= -\{[(1/2) \times (L/2) \times (+WL/4EI) \times (2/3) \times (L/2)] + [(L/2) \times (-M/EI) \times (L/4)]\}$$

Substituting $M = WL/8$ and rearranging

$$y_C = +WL^3/(192EI)$$

The positive sign means the deflection is downwards, which is obviously correct. Notice that by choosing the origin at A the left hand side of the equation reduces to the deflection at C.

3.3 Moment applied at the end of a fixed ended beam

> If a moment M is applied at the pinned support of a propped cantilever (see Fig. 3.3(a)), determine the moment transferred to the fixed end.

Solution Draw the M/EI diagrams shown in Figs 3.3(b) and (c). The signs of the bending moments are known, or assumed, and are in accordance with the right hand screw rule. For convenience the diagrams are reduced to simple cases. Figure 3.3(b) is the diagram for the moment M applied at the end of a simply supported beam. Figure 3.3(c) is the diagram for the induced end moment M_A for a simply supported beam.

To determine the moment induced at A, apply equation (3.3) between A and B with the origin at B:

$$[z \, dy/dz - y]_0^L = -[A\bar{z}/EI]_0^L$$

and

$$[L \times 0 - 0] - [0 \times \theta_B - 0] = -\{[(1/2) \times L \times (+M_A/EI) \times (2/3) \times L] + [(1/2) \times L \times (-M/EI) \times (L/3)]\}$$

Hence

$$M_A = +M/2$$

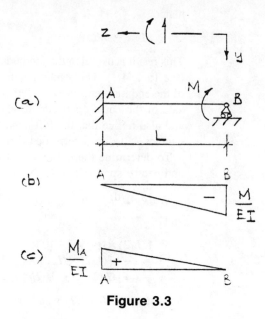

Figure 3.3

This result is important for use in the method of moment distribution (see Chapter 5).

The positive sign means that the assumed direction of the moment M_A is correct. Notice that by choosing the origin at B the left hand side of the equation reduces to zero.

3.4 End moments induced by settlement

If the supports of a fixed ended beam move relative to each other by a distance y_B, determine the fixed end moments (see Fig. 3.4(a)).

Solution Draw the M/EI diagrams shown in Figs 3.4(b) and (c). The signs of the bending moments are known, or assumed, and are in accordance with the right hand screw rule. For convenience the bending moment diagrams are reduced to simple cases. Figures 3.4(b) and (c) are diagrams for end moments induced by settlement at the ends of a cantilever beam.

To determine the induced end moments, apply equation (3.3) between A and B with the origin at A:

$$[z \, dy/dz - y]_0^L = -[A\bar{z}/EI]_0^L$$

and

$$[0 - y_B] - [0 \times 0 - 0] = -\{ [(1/2) \times L \times (+M/EI) \times (2/3) \times L] + [(1/2) \times L \times (-M/EI) \times (L/3)] \}$$

Hence

$$M = +6EIy_B/L^2$$

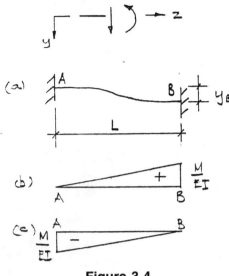

Figure 3.4

This result is useful in the methods of moment distribution and slope deflection (see Table 4.1).

3.5 Propped cantilever

> Determine the deflection at the end of the propped cantilever shown in Fig. 3.5(a).

Solution   Separate the *M/EI* diagrams into simple cases as shown in Figs 3.5(b) and (c). Figure 3.5(b) is for the cantilever load on a simply supported span and Fig. 3.5(c) is for the end moment at A on a simply supported beam, see Example 3.3. The signs of the bending moments are known, or assumed, and are in accordance with the right hand screw rule.

For the cantilever BC

$$\text{B} \quad +M_{BC}+w\times(L/2)\times(L/4) = 0; \quad \text{hence } M_{BC} = -wL^2/8$$

To determine the slope at B, apply equation (3.2) between A and B with the origin at A:

$$[dy/dz]_0^L = -[A/EI]_0^L$$

and

$$[\theta_B - 0] = -\{[(1/2)\times L\times(+M_B/2EI)]+[(1/2)\times L\times(-M_B/EI)]\}$$

Hence

$$\theta_B = +M_B L/(EI)$$

Since

$$M_B = M_{BC}, \; \theta_B = +wL^3/(8EI) \tag{1}$$

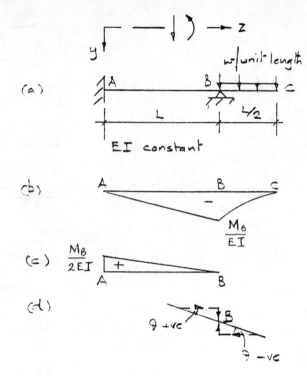

Figure 3.5

To determine the deflection at C, apply equation (3.3) between B and C with the origin at C:

$$[z \, dy/dz - y]_0^{L/2} = -[A\bar{z}/EI]_0^{L/2}$$

and

$$[(L/2) \times \theta_B - 0] - [0 \times \theta_C - y_C] =$$
$$-\{[(1/3) \times (L/2) \times (-M_B/EI) \times (3/4) \times (L/2)]\}$$

Rearranging

$$(L/2) \times \theta_B + y_C = +M_B L^2/(16EI) \tag{2}$$

Combining (1) and (2) and rearranging

$$y_C = +9wL^4/(128EI)$$

Notice the change in sign of θ_B as the origin changes position (see Fig. 3.5(d)).

3.6 Distribution of a moment applied at a joint

If a moment M is applied at joint B of the two span beam shown in Fig. 3.6(a), determine the distribution of moments.

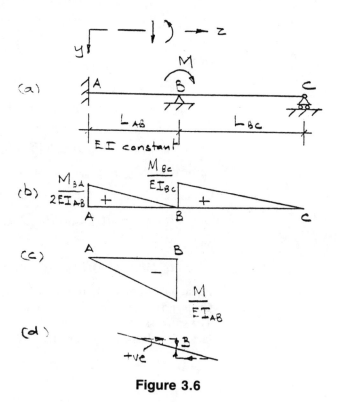

Figure 3.6

Solution Draw the M/EI diagrams shown in Figs 3.6(b) and (c). Reduce the diagrams to cases for simply supported beams as shown in Figs 3.6(b) and (c). The signs of the bending moments are known, or assumed, and are in accordance with the right hand screw rule.

To determine the slope at B, apply equation (3.2) between A and B with the origin at A:

$$[dy/dz]_0^{L_{AB}} = -[A/EI]_0^{L_{AB}}$$

and

$$[\theta_B - 0] = -\{[(1/2) \times L_{AB} \times (+M_{BA}/2EI_{AB})] + [(1/2) \times L_{AB} \times (-M_{BA}/EI_{AB})]\}$$

Hence

$$\theta_B = +M_{BA}L_{AB}/(4EI_{AB}) \tag{1}$$

The angle is positive as indicated in Fig. 3.6(d).

An alternative expression for the slope at B is obtained by applying equation (3.3) between B and C with the origin at B:

$$[z \, dy/dz - y]_0^{L_{BC}} = -[A\bar{z}/EI]_0^{L_{BC}}$$

and

$$[L_{BC} \times \theta_B - 0] - [0 \times \theta_C - 0] = -[(1/2) \times L_{BC} \times (+M_{BC}/EI_{BC}) \times (2/3) \times L_{BC}]$$

Hence

$$\theta_B = -M_{BC}L_{BC}/(3EI_{BC}) \tag{2}$$

The angle is negative as indicated in Fig. 3.6(d). Equating the two values of θ_B given in equations (1) and (2)

$$M_{BC}/M_{BA} = 3EI_{BC}L_{AB}/(4EI_{AB}L_{BC}) \tag{3}$$

Also $M = M_{BA} + M_{BC}$ (4)

Combining equations (3) and (4) and rearranging

$$M_{BA}/M = (EI_{AB}/L_{AB})/[EI_{AB}/L_{AB} + (3/4)EI_{BC}/L_{BC}] \tag{5}$$

and

$$M_{BC}/M = (3/4)(EI_{BC}/L_{BC})/[EI_{AB}/L_{AB} + (3/4)EI_{BC}/L_{BC}] \tag{6}$$

Equations (5) and (6) are useful for the method of moment distribution (see Chapter 5). The right hand side of the equation is called the distribution factor. The stiffness of a member is defined as kEI/L. For member AB, the stiffness factor $k = 1$ and for member BC, $k = 3/4$.

3.7 Distribution of a moment applied at a joint

Determine the distribution of moments for the three span beam shown in Fig. 3.7(a).

Solution Draw the M/EI diagrams shown in Figs 3.7(b and c). Reduce the diagrams to cases for simply supported beams as shown in Fig. 3.7(b). Figure 3.7(c) is an alternative way of expressing the variation in bending moment. The signs of the bending moments are known, or assumed, and are in accordance with the right hand screw rule.

From the skew-symmetry of the deflected structure, shown dotted in Fig. 3.7(a), there is effectively a pin at E. From Example 3.3

$$M_{AB} = (1/2)M_{BA}$$

To determine the slope at B, apply equation (3.2) between A and B with the origin at A:

$$[dy/dz]_0^{L_{AB}} = -[A/EI]_0^{L_{AB}}$$

and

$$[\theta_B - 0] = -[(1/2) \times L_{AB} \times (+M_{BA}/2EI_{AB}) + (1/2) \times L_{AB} \times (-M_{BA}/2EI_{AB})]$$

Hence

$$\theta_B = -M_{BA}L_{AB}/(4EI_{AB}) \tag{1}$$

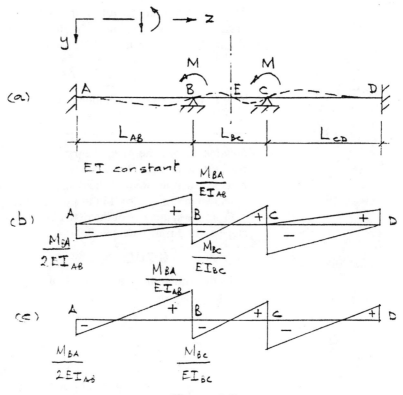

Figure 3.7

An alternative expression for the slope at B is obtained by applying equation (3.3) between B and E with the origin at E:

$$[z \, dy/dz - y]_0^{L_{BC}/2} = -[A\bar{z}/EI]_0^{L_{BC}/2}$$

and

$$[(L_{BC}/2) \times \theta_B - 0] - [0 \times \theta_E - 0] = -[(1/2) \times (L_{BC}/2) \times (-M_{BC}/EI_{BC}) \times (2/3) \times (L_{BC}/2)]$$

Hence

$$\theta_B = +M_{BC}L_{BC}/(6EI_{BC}) \tag{2}$$

Equating the two values of θ_B from equations (1) and (2)

$$M_{BC}/M_{BA} = 6EI_{BC}L_{AB}/(4EI_{AB}L_{BC}) \tag{3}$$

Also $M = M_{BA} + M_{BC}$ $\qquad\qquad$ (4)

Combining equations (3) and (4)

$$M_{BA}/M = (EI_{AB}/L_{AB})/[EI_{AB}/L_{AB} + (3/2)EI_{BC}/L_{BC}]$$

and

$$M_{BC}/M = (EI_{BC}/L_{BC})/[EI_{AB}/L_{AB} + (3/2)EI_{BC}/L_{BC}]$$

These equations are useful for the method of moment distribution. The right hand side of the equation is the distribution factor. The stiffness factor for BC in this situation is $k = 3/2$ (see Table 5.1 and Example 5.12).

3.8 Distribution of moments applied at joints

Determine the distribution of moments for the three span beam shown in Fig. 3.8(a).

Solution Draw the M/EI diagrams shown in Fig. 3.8(b). The signs of the bending moments are known, or assumed, and are in accordance with the right hand screw rule. Reduce the diagrams to cases for simply supported beams as shown in Fig. 3.8(b). In this problem the diagrams are plotted on one axis.

From the symmetry of the deflected structure there is a deflection and zero slope at E. From Example 3.3

$$M_{AB} = (1/2)M_{BA}$$

To determine the slope at B, apply equation (3.2) between A and B with the origin at A:

$$[dy/dz]_0^{L_{AB}} = -[A/EI]_0^{L_{AB}}$$

and

$$[\theta_B - 0] = -[(1/2) \times L_{AB} \times (+M_{BA}/EI_{AB}) + (1/2) \times L_{AB} \times (-M_{BA}/2EI_{AB})]$$

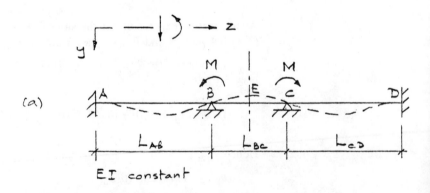

(a)

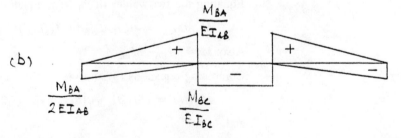

(b)

Figure 3.8

Hence

$$\theta_B = -M_{BA}L_{AB}/(4EI_{AB}) \tag{1}$$

An alternative expression for the slope at B is obtained by applying equation (3.3) between B and C with the origin at C, giving

$$[z \, dy/dz - y]_0^{L_{BC}} = -[A\bar{z}/EI]_0^{L_{BC}}$$

and

$$[L_{BC} \times \theta_B - 0] - [0 \times \theta_C - 0] = -[L_{BC} \times (-M_{BC}/EI_{BC}) \times (L_{BC}/2)]$$

Hence

$$\theta_B = +M_{BC}L_{BC}/(2EI_{BC}) \tag{2}$$

Equating the two values of θ_B from equations (1) and (2)

$$M_{BC}/M_{BA} = EI_{BC}L_{AB}/(2EI_{AB}L_{BC}) \tag{3}$$

Also $M = M_{BA} + M_{BC}$ \tag{4}

Combining equations (3) and (4)

$$M_{BA}/M = (EI_{AB}/L_{AB})/[EI_{AB}/L_{AB} + (1/2)EI_{BC}/L_{BC}]$$

and

$$M_{BC}/M = (EI_{BC}/L_{BC})/[EI_{AB}/L_{AB} + (1/2)EI_{BC}/L_{BC}]$$

These equations are useful for the method of moment distribution. The right hand side of the equation is the distribution factor and the stiffness factor for BC in this situation is $k = 1/2$ (see Chapter 5, Example 5.8).

3.9 Distribution of moments applied at joints

Determine the distribution of moments for the portal frame shown in Fig. 3.9(a).

Solution Draw the *M/EI* diagrams shown in Fig. 3.9(b). The signs of the bending moments are known, or assumed, and are in accordance with the right hand screw rule. Reduce the diagrams to cases for simply supported members as shown in Fig. 3.9(b).

From the symmetry of the deflected structure there is an effective pin at E with zero vertical deflection.

To determine the slope at B, apply equation (3.2) between A and B with the origin at A:

$$[dy/dz]_0^{L_{AB}} = -[A/EI]_0^{L_{AB}}$$

and

$$[\theta_B - 0] = -[L_{AB} \times (+M_{BA}/EI_{AB})]$$

Hence

$$\theta_B = -M_{BA}L_{AB}/(EI_{AB}) \tag{1}$$

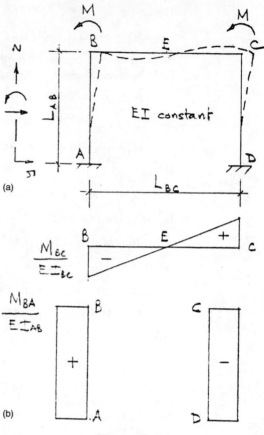

Figure 3.9

An alternative expression for the slope at B is obtained by applying equation (3.3) between B and E with the origin at E:

$$[z\, dy/dz - y]_0^{L_{BC}/2} = -[A\bar{z}/EI]_0^{L_{BC}/2}$$

$$[(L_{BC}/2) \times \theta_B - 0] - [0 \times \theta_E - 0] = -[(1/2) \times (L_{BC}/2) \times (-M_{BC}/EI_{BC}) \times (2/3) \times (L_{BC}/2)]$$

Hence

$$\theta_B = +M_{BC}L_{BC}/(6EI_{BC}) \qquad (2)$$

Equating the two values of θ_B from equations (1) and (2)

$$M_{BC}/M_{BA} = 6EI_{BC}L_{AB}/(EI_{AB}L_{BC}) \qquad (3)$$

Also $M = M_{BA} + M_{BC}$ (4)

Combining equations (3) and (4)

$$M_{BA}/M = (EI_{AB}/L_{AB})/[EI_{AB}/L_{AB} + 6EI_{BC}/L_{BC}]$$

and

$$M_{BC}/M = (EI_{BC}/L_{BC})/[EI_{AB}/L_{AB}+6EI_{BC}/L_{BC}]$$

These equations are also useful for the method of moment distribution (see Chapter 5, Example 5.11). Notice also that $M_{AB} = M_{BA}$ which explains the carry over factor of -1 used in moment distribution.

3.10 Variable second moment of area

Determine the slope and deflection at C for the mast with the variable second moment of area shown in Fig. 3.10(a).

Solution Draw the M/EI diagrams shown in Fig. 3.10(b). The signs of the bending moments are known, or assumed, and are in accordance with the right hand screw rule. The bending moment diagram in Fig. 3.10(c) is stepped because of the variation in the second moment of area.

To determine the slope at C, apply equation (3.2) between A and C with the origin at A:

$$[dy/dz]_0^{2H} = -[A/EI]_0^{2H}$$

and

$$[\theta_C - 0] = -[(1/2) \times H \times (+W_BH/2EI) + (1/2) \times H \times (+W_CH/EI)$$
$$+ H \times (+W_CH/2EI) + (1/2) \times H \times (+W_CH/2EI)]$$

Hence

$$\theta_C = -H^2(W_B + 5W_C)/(4EI)$$

To determine the deflection at C apply equation (3.3) between A and C with the origin at C:

$$[z \, dy/dz - y]_0^{2H} = -[A\bar{z}/EI]_0^{2H}$$

and

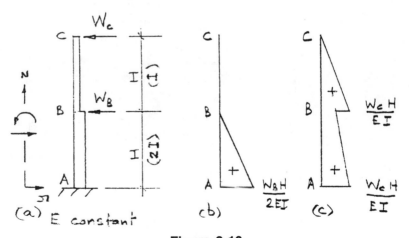

Figure 3.10

$$[2H \times 0 - 0] - [0 \times \theta_C - y_C] = -[(1/2) \times H \times (+W_BH/2EI) \times (H + 2H/3)$$
$$+ (1/2) \times H \times (+W_CH/EI) \times (2H/3) + H \times (+W_CH/2EI) \times (3H/2)$$
$$+ (1/2) \times H \times (+W_CH/2EI) \times (H + 2H/3)]$$

Hence

$$y_C = -H^3(5W_B + 18W_C)/(12EI)$$

3.11 Sway frame

Determine the slope and moment at B for the sway frame shown in Fig. 3.11(a).

Solution Draw the *M/EI* diagrams shown in Fig. 3.11(b). The signs of the bending moments are known, or assumed, and are in accordance with the right hand screw rule.

To determine the slope at B, apply equation (3.2) between A and B with the origin at A:

$$[dy/dz]_0^H = -[A/EI]_0^H$$

and

$$[\theta_B - 0] = -[(1/2) \times H \times (+M_B/EI) + (1/2) \times H \times (-M_A/EI)]$$

Hence

$$\theta_B = -H(M_B - M_A)/(2EI) \qquad (1)$$

An alternative expression for the slope at B is obtained by applying equation (3.3) between B and C with the origin at C, giving

$$[z \, dy/dz - y]_0^L = -[A\bar{z}/EI]_0^L$$

and

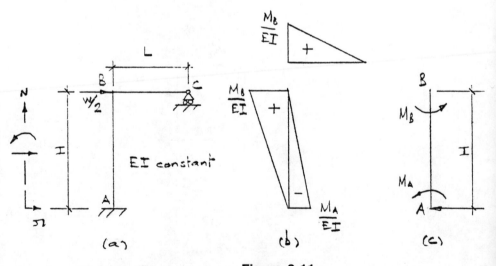

Figure 3.11

(a)   (b)   (c)

$$[L \times \theta_B - 0] - [0 \times \theta_C - 0] = -[(1/2) \times L \times (+M_B/EI) \times (2/3) \times L]$$

Hence

$$\theta_B = -M_B L/(3EI) \qquad (2)$$

Equating the two values of θ_B from equations (1) and (2)

$$M_A/M_B = [(2/3) \times (L/H) + 1] \qquad (3)$$

For the equilibrium of member AB (see Fig. 3.11(c))

$$B) \quad -M_B - M_A + WH/2 = 0 \qquad (4)$$

Combining equations (3) and (4) and rearranging

$$M_B = (3WH)/(4L/H + 12) \qquad (5)$$

Substituting M_B from (5) in (4) and rearranging

$$\theta_B = (WHL/EI)/(4L/H + 12)$$

3.12 Derivation of slope deflection equations

A typical loaded member AB is shown in Fig. 3.12(a). The member is shown displaced from its original position by end rotations θ_{AB} and θ_{BA} and a relative end displacement y. Form general expressions for the end moments M_{AB} and M_{BA} in terms of the bending stiffness EI/L, the end rotations θ_{AB} and θ_{BA}, the displacement y, and the fixed end moments.

Solution Break down the effects of loading and displacements into separate cases as shown in Figs 3.12(b) to (e). Draw the *M/EI* diagrams for each case and use the result obtained from Example 3.3 that if a moment is applied at one end of a member half that moment is induced at the fixed end. The signs of the bending moments are known and are in accordance with the right hand screw rule.

End moment at A

$$M_{AB} = M_1 + M_2/2 + M_3 + M_4 \qquad (1)$$

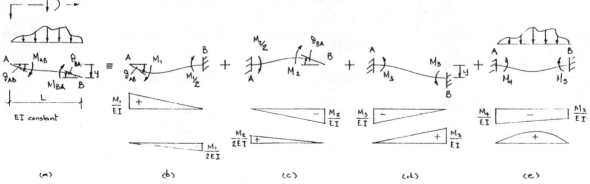

(a) (b) (c) (d) (e)

Figure 3.12

To determine M_1, apply equation (3.2) between A and B with the origin at A (see Fig. 3.12(b)):

$$[dy/dz]_0^L = -[A/EI]_0^L$$

and

$$[0-\theta_{AB}] = -[(1/2)\times L\times(+M_1/EI)+(1/2)\times L\times(-M_1/2EI)]$$

Hence
$$M_1 = +4EI\theta_{AB}/L \qquad (2)$$

To determine M_2, apply equation (3.2) between A and B with the origin at A (see Fig. 3.12(c)):

$$[dy/dz]_0^L = -[A/EI]_0^L$$

and

$$[\theta_{BA}-0] = -[(1/2)\times L\times(-M_2/EI)+(1/2)\times L\times(+M_2/2EI)]$$

Hence

$$M_2 = +3EI\theta_{BA}/L \qquad (3)$$

To determine M_3, apply equation (3.3) between A and B with the origin at A (see Fig. 3.12(d) and Example 3.4):

$$M_3 = -6EIy/L^2 \qquad (4)$$

M_4 is a fixed end moment (FEM) and depends on the loading conditions. Apply equation (3.2) between A and B with the origin at A (see Fig. 3.12(e)). A typical case is shown in Example 3.2. Expressed generally

$$M_4 = \pm FEM \qquad (5)$$

Inserting equations (2) to (5) in (1)

$$M_{AB} = (2EI/L)(2\theta_{AB}+\theta_{BA}-3y/L)\pm FEM$$

Similarly

$$M_{BA} = (2EI/L)(2\theta_{BA}+\theta_{AB}-3y/L)\pm FEM$$

These are important equations and are applied to problems using the method of slope deflection in Chapter 4.

Problems

1 Determine the slope and deflection at the end of the cantilever shown in Fig. 3.13.

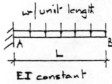

Figure 3.13

Solution

Apply equation (3.2) between A and B with the origin at A, hence
$\theta_B = wL^3/(6EI)$.

Apply equation (3.3) between A and B with the origin at B, hence
$y_B = wL^4/(8EI)$.

2 Determine the fixed end moments for the beam shown in Fig. 3.14.

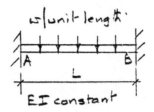

Figure 3.14

Solution

Apply equation (3.2) between A and B with the origin at A, hence
$\text{FEM}_A = -wL^2/12$, $\text{FEM}_B + wL^2/12$.

3 Determine the fixed end moment at A for the deflected propped cantilever shown in Fig. 3.15.

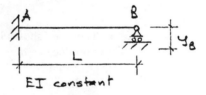

Figure 3.15

Solution

Apply equation (3.3) between A and B with the origin at B, hence
$\text{FEM}_A = -3EIy_B/L^2$.

4 Determine the deflection at mid span of the beam shown in Fig. 3.16 if a moment M is applied at support A.

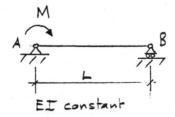

Figure 3.16

Apply equation (3.3) between A and B with the origin at A, hence
$\theta_B = -ML/(6EI)$.

Apply equation (3.3) between centre span and B with the origin at centre span, hence $y = ML^2/(16EI)$.

5 Determine the slope and deflection at the end of the cantilever shown in Fig. 3.17.

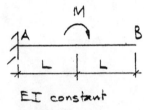

EI constant

Figure 3.17

Solution

Apply equation (3.2) between A and B with the origin at A, hence
$\theta_B = ML/(EI)$.

Apply equation (3.3) between A and B with the origin at A, hence
$y_B = 3ML^2/(2EI)$.

6 The beam shown in Fig. 3.18 is of constant *EI* value and the loading at C produces a deflection of 5 mm at a point midway between A and B. Determine

(a) the *EI* value

(b) the value of the maximum deflection between A and B

(c) the value of the deflection at C.

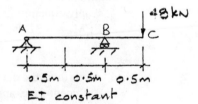

EI constant

Figure 3.18

Solution

(a) Apply equation (3.3) between A and B with the origin at A, hence
$\theta_B = +8/(EI)$.

Apply equation (3.3) between D and B with the origin at D, hence
$EI = 300 \text{ Nm}^2$

(b) Apply equation (3.2) between A and B with the origin at A, hence $\theta_A = -4/(EI)$.

Apply equation (3.2) between A and Z with the origin at A, hence $\theta_A = 12z^2/(EI)$.

Equating values of θ_A position of maximum deflection is $1/\sqrt{3}$ m from A.

Apply equation (3.3) between A and Z with the origin at A, hence $y_{max} = 80\sqrt{3}/27$ mm.

(c) Apply equation (3.3) between C and B with the origin at C, and insert the value of θ_B from (a), hence $y_C = 20$ mm.

7 The beam shown in Fig. 3.19 is of constant EI value and is loaded with loads W_1 and W_2. Determine the ratio of W_1/W_2 if the deflection at C is zero.

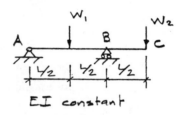

Figure 3.19

Solution

Apply equation (3.3) between A and B with the origin at A, hence $\theta_B = (W_2/6 - W_1/16)L^2/(EI)$.

Apply equation (3.3) between C and B with the origin at C, hence $W_1/W_2 = 4$.

8 The cantilever shown in Fig. 3.20 is of variable EI value. Determine the slope and deflection at the end of the cantilever.

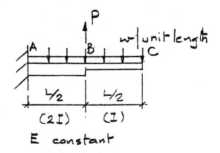

Figure 3.20

Apply equation (3.2) between A and C with the origin at A, hence $\theta_C = 3wL^3/(32EI) - PL^2/(16EI)$.

Apply equation (3.3) between C and A with the origin at C, hence $y_C = 17wL^4/(256EI) - 5PL^3/(96EI)$.

9 The bent cantilever shown in Fig. 3.21 is of constant EI value. Determine the slope and deflection at the end of the cantilever.

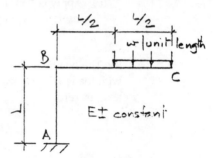

Figure 3.21

Solution

Apply equation (3.2) between A and B with the origin at A, hence $\theta_B = +3wL^3/(8EI)$.

Apply equation (3.3) between B and A with the origin at B, hence $y_B = 3wL^3/(16EI)$.

Apply equation (3.3) between C and B with the origin at C, and insert value of θ_B with a change in sign, hence $y_C = +1427wL^4/(3072EI)$.

10 The two span continuous beam shown in Fig. 3.22 is of variable EI value. Determine the moments at A and B.

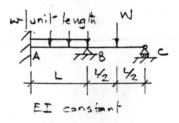

Figure 3.22

Solution

Apply equation (3.3) between C and B with the origin at C to produce an equation in θ_B.

Apply equation (3.2) between B and A with the origin at B to produce another equation in θ_B

Equating to form $M_A + 5M_B/3 = wL^2/6 + WL/8$ (i)

Apply equation (3.3) between B and A with the origin at B to form
$M_A + M_B/2 = wL^2/8$ (ii)

Solving equations (i) and (ii),

$$M_A = 3(wL^2 - WL/2)/28, \quad M_B = (wL^2 + 3WL)/28.$$

4

Slope deflection method

Introduction

Slope deflection is a method of structural analysis based on the stiffness method which is applicable to certain types of elastic hyperstatic structures where members are predominantly in bending. The method is generally limited to structures in one plane and assumes that axial and shear forces have little effect on the solution of a structure.

The method results in a series of simultaneous equations which involve end moments, deflections and slopes. If the number of equations is small they can be solved by hand calculation but if numerous they are solved using a computer.

The method does not require the calculation of areas of bending moment diagrams as in area-moment, but it is not such a good method for understanding the basic principles. The solution also produces values of slopes which are not always required in design practice.

The first step in the analysis of a hyperstatic structure using slope deflection is to split the structure into fixed ended beam cases, e.g. the continuous two span beam shown in Fig. 4.1(a) is split into two fixed ended beams (see Fig. 4.1(b)). The fixed end moments (FEMs) for the load case on each span are determined from Table 4.1. These values have been calculated using Mohr area and area-moment methods as shown in Examples 3.2 and 3.4. The fixed end moments are shown for a variety of end conditions, relative displacements of the supports and various types of loading.

The fixed end moments for each member are adjusted to the real structure by corrective rotations and deflections at the ends of each member. If these corrections are carried out theoretically for a typical member AB (see Fig. 4.2(a)) using area and area-moment methods, then the following equations result (see Example 3.12).

If $y = 0$ then this produces the following flexibility equations relating the rotation at the ends to the end moments.

$$\phi_{AB} = [L/(6EI)](2M_{AB} - M_{BA}) \tag{4.1}$$

$$\phi_{BA} = [L/(6EI)](-M_{AB} + 2M_{BA}) \tag{4.2}$$

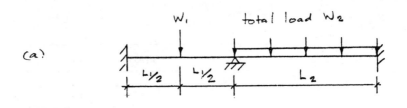

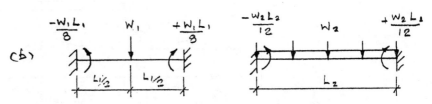

Figure 4.1

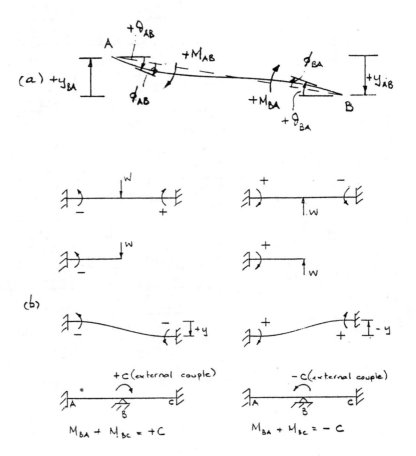

Figure 4.2

Table 4.1

Fixed end moments		
$+WL/8$	$-WL/12$ total W $+WL/12$ L	$-WL/8$
$+5WL/32$	$-5WL/48$ total W $+5WL/48$	$-5WL/32$
$+2WL/15$	$-WL/15$ total W $+WL/10$	$-7WL/60$
$+\dfrac{Wa}{8}\left(2-\dfrac{a}{L}^2\right)$	$-\dfrac{Wa}{12}\left[6-8\dfrac{a}{L}+3\left(\dfrac{a}{L}\right)^2\right]$ total W $+\dfrac{Wa^2}{12L}\left(4-\dfrac{3a}{L}\right)$ a	$-\dfrac{Wa}{8}\left(2-\dfrac{a}{L}\right)^2$
W $+3WL/16$	$-WL/8$ W $+WL/8$ $L/2$	$-3WL/16$ W
W $+\dfrac{Wab(2a+b)}{2L^2}$	$-\dfrac{Wab^2}{L}$ W $+\dfrac{Wa^2b}{L^2}$ a b	$-\dfrac{Wab(a+2b)}{2L^2}$ W
$W/2$ $W/2$ $+WL/6$	$-WL/9$ $W/2$ $W/2$ $+WL/9$ $L/3$ $L/3$ $L/3$	$-WL/6$ $W/2$ $W/2$
$-\dfrac{M}{2}\left[2-6\dfrac{b}{L}+3\left(\dfrac{b}{L}\right)^2\right]$ $+M$	$+\dfrac{Mb}{L^2}(3a-L)$ $+M$ $+\dfrac{Ma}{L^2}(3b-L)$ a b	$-\dfrac{M}{2}\left[2-6\dfrac{a}{L}+3\left(\dfrac{a}{L}\right)^2\right]$ $+M$
$+3EI\Delta/L^2$	$-6EI\Delta/L^2$ $-6EI\Delta/L^2$ Δ	$-3EI\Delta/L^2$
$-3EI\Delta/L^2$	$+6EI\Delta/L^2$ $+6EI\Delta/L^2$ Δ	$+3EI\Delta/L^2$

These equations can be modified for $y \neq 0$ and the introduction of fixed end moments. Solving for M_{AB} and M_{BA} and replacing ϕ by $(\theta - y/L)$

$$M_{AB} = (2EI/L)(2\theta_A + \theta_B - 3y/L) \pm \text{FEM} \qquad (4.3)$$

$$M_{BA} = (2EI/L)(2\theta_B + \theta_A - 3y/L) \pm \text{FEM} \qquad (4.4)$$

where θ is a rotation at the end of a member, y is the deflection at right angles to a member relative to the other end (see Fig. 4.2), L is the actual length of a member between intersections (nodes or joints), E is Young's modulus of elasticity, and I is the second moment of area of member generally assumed to be constant between intersections (nodes or joints).

The sign convention is based on the right hand screw rule as described in Volume 1 of *Solving Problems in Structures*. Figure 4.2(a) shows positive clockwise for moments, rotations and deflections for the directions shown and examples of signs in other situations are given in Fig. 4.2(b). Axial force, shear force and bending moment diagrams are also constructed based on the right hand screw rule. The signs shown in equations (4.3) and (4.4) are consistent with the displaced shape of the member shown in Fig. 4.2(a). In problems where the sign of a moment, rotation, or deflection is known then the correct sign is inserted in the equation. Where it is not known the signs in the equations remain unaltered as shown in the following problems.

Worked examples

4.1 Propped cantilever

> Determine the slope at B and draw the bending moment diagram for the beam shown in Fig. 4.3(a).

Solution This is a simple hyperstatic structure with one redundancy. From Table 4.1 the fixed end moments for span AB are

$$\text{FEM}_{AB} = \text{FEM}_{BA} = WL/8$$

Form the following equations using the slope deflection sign convention. Where signs are not known then leave as in equations (4.3) and (4.4), e.g. θ_B. Other values (θ_A and y) are known to be zero. The directions of the fixed end moments are known and given the correct sign.
From equation (4.3) applied to span AB

$$M_{AB} = (2EI/L_{AB})(0 + \theta_B + 0) - WL_{AB}/8 \qquad (1)$$

From equation (4.4) applied to span AB

$$M_{BA} = (2EI/L_{AB})(2\theta_B + 0 + 0) + WL_{AB}/8 \qquad (2)$$

For the cantilever member BC

$$M_{BC} = -wL_{BC}^2/2 \qquad (3)$$

At joint B the moments balance

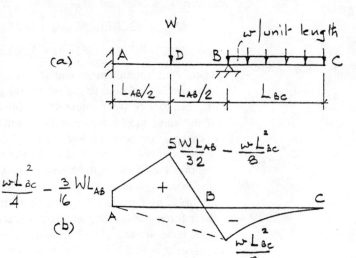

Figure 4.3

$$B) + M_{BA} + M_{BC} = 0 \tag{4}$$

There are four equations and four unknowns (M_{AB}, M_{BA}, M_{BC} and θ_B) and these can now be solved by hand calculation. Generally it is less complicated to determine the slopes first then the moments. The first quantity to be determined is the most difficult.

Substituting (2) and (3) in (4) and rearranging

$$\theta_B = (L_{AB}/4EI)(wL_{BC}^2/2 - WL_{AB}/8) \tag{5}$$

Note that if $wL_{BC}^2/2 > WL_{AB}/8$ then θ_B is clockwise positive, and if $wL_{BC}^2/2 < WL_{AB}/8$ then θ_B is anti-clockwise negative.

Substituting (5) in (1)

$$M_{AB} = wL_{BC}^2/4 - 3WL_{AB}/16 \tag{6}$$

Note the change in sign of the bending moment M_{AB} depending on the magnitudes of the quantities involved. This sign change is obvious if the structure and effects of the loads are considered, e.g. if $W = 0$ then M_{AB} is positive.

The bending moment diagram is shown in Fig. 4.3(b). The bending moment at D is obtained from adjusting the simply supported bending moment by the end bending moments (see Fig. 4.3(b)):

$$
\begin{aligned}
M_D &= WL_{AB}/4 + (M_{AB} + M_{BC})/2 \\
&= WL_{AB}/4 + (wL_{BC}^2/4 - 3WL_{AB}/16)/2 - wL_{BC}^2/2 \\
&= 5WL_{AB}/32 - wL_{BC}^2/8
\end{aligned}
$$

4.2 Two span beam

Determine the slopes at B and C and draw the bending moment diagram for the continuous two span beam shown in Fig. 4.4(a). This is a more complicated hyperstatic structure with two redundancies and consequently the number of equations to solve is increased to 5.

Solution From Table 4.1 the fixed end moments are

$$\text{FEM}_{AB} = \text{FEM}_{BA} = wL^2/12$$
$$\text{FEM}_{BC} = \text{FEM}_{CB} = WL/8$$

Form the following equations using the slope deflection sign convention. Where signs are not known then they are assumed to be positive, and the fixed end moments are given the appropriate sign.
From equation (4.3) applied to span AB

$$M_{AB} = (2EI/L)(0 + \theta_B + 0) - wL^2/12 \tag{1}$$

From equation (4.4) applied to span AB

$$M_{BA} = (2EI/L)(2\theta_B + 0 + 0) + wL^2/12 \tag{2}$$

From equation (4.3) applied to span BC

$$M_{BC} = (2EI/L)(2\theta_B + \theta_C + 0) - WL/8 \tag{3}$$

From equation (4.4) applied to span BC

$$M_{CB} = 0 = (2EI/L)(2\theta_C + \theta_B + 0) + WL/8 \tag{4}$$

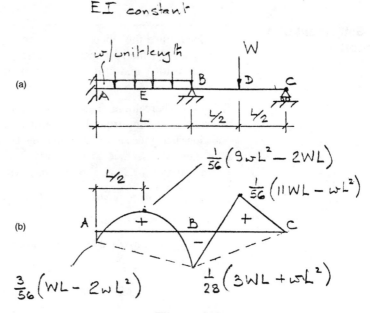

(a)

(b)

Figure 4.4

At joint B

$$B) +M_{BA}+M_{BC} = 0 \tag{5}$$

There are five equations and five unknowns (M_{AB}, M_{BA}, M_{BC}, θ_B and θ_C), which can now be solved. First determine the slopes θ_B and θ_C then the bending moments.

Substituting (2) and (3) in (5) and combining with (4)

$$\theta_B = (L/28EI)(3WL/4 - wL^2/3) \tag{6}$$

and from (4)

$$\theta_C = (L/28EI)(wL^2/6 - 5WL/4) \tag{7}$$

Substituting (6) in (1)

$$M_{AB} = 3(WL/2 - wL^2)/28$$

Substituting (6) in (2)

$$M_{BA} = (3WL + wL^2)/28$$

The bending moment diagram is shown in Fig. 4.4(b). The bending moments at D and E are obtained from adjusting the simply supported bending moment by the end bending moments (see Fig. 4.4(b)):

$$M_D = WL/4 + M_{BC}/2 = WL/4 - (3WL + wL^2)/56 = (11WL - wL^2)/56$$

and

$$\begin{aligned} M_E &= wL^2/8 + (M_{AB} + M_{BA})/2 \\ &= wL^2/8 - 3(WL/2 - wL^2)/56 - (WL/2 + wL^2)/56 \\ &= (9wL^2 - 2WL)/56 \end{aligned}$$

4.3 Settlement of a support

Determine the slopes at A, B, and C and draw the bending moment diagram for the two span beam shown in Fig. 4.5(a). Support B sinks relative to A and C introduces bending moments into the structure.

Solution There are no fixed end moments from loads in this problem but there are values resulting from the settlement of the support B. Form the following equations using the slope deflection sign convention. Where signs are not known then they are assumed to be positive.

From equation (4.3) applied to span AB

$$M_{AB} = 0 = (2EI/L)(2\theta_A + \theta_B - 3y_B/L) + 0 \tag{1}$$

From equation (4.4) applied to span AB

$$M_{BA} = (2EI/L)(2\theta_B + \theta_A - 3y_B/L) + 0 \tag{2}$$

From equation (4.3) applied to span BC

$$M_{BC} = (2EI/2L)(2\theta_B + \theta_C + 3y_B/2L) + 0 \tag{3}$$

From equation (4.4) applied to span BC

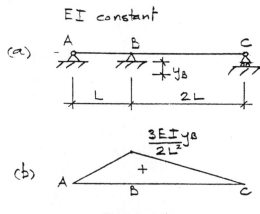

EI constant

(a)

(b)

Figure 4.5

$$M_{CB} = 0 = (2EI/2L)(2\theta_C + \theta_B + 3y_B/2L) + 0 \qquad (4)$$

The signs for the deflection (y_B) in (3) and (4) are changed because the deflected shape of BC is the opposite to that shown in Fig. 4.2(a).
At joint B

$$B) \quad +M_{BA} + M_{BC} = 0 \qquad (5)$$

There are five equations and five unknowns $(M_{BA}, M_{BC}, \theta_A, \theta_B$ and $\theta_C)$. Substituting (2) and (3) in (5) and combining with (1) and (4)

$$\theta_A = +5y_B/4L \text{ (clockwise positive)} \qquad (6)$$

and from (1)

$$\theta_B = +y_B/2L \text{ (clockwise positive)} \qquad (7)$$

and from (4)

$$\theta_C = -y_B/L \text{ (anti-clockwise negative)} \qquad (8)$$

Substituting (7) and (8) in (3)

$$M_{BC} = 3EIy_B/(2L^2)$$

The bending moment diagram is shown in Fig. 4.5(b).

4.4 Rectangular sway frame

Determine the slopes at B and C, the deflection at B, and draw the bending moment diagram for the frame shown in Fig. 4.6(a). The frame sways horizontally by an unknown amount y_B which introduces additional bending moments into the structure.

Solution There are no fixed end moments because the load $(W/2)$ is applied at the joint, but the horizontal deflection of B induces end moments. Form the following equations using the slope deflection sign convention. Where signs are not known then they are assumed to be positive. If sway to the right is

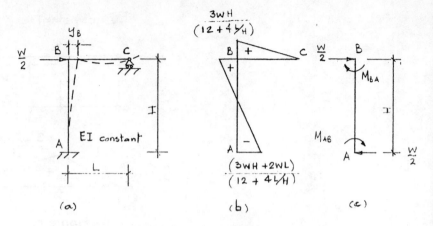

Figure 4.6

assumed then the deflected shape of AB agrees with that of Fig. 4.2(a).
From equation (4.3) applied to member AB

$$M_{AB} = (2EI/H)(0+\theta_B-3y_B/H)+0 \qquad (1)$$

From equation (4.4) applied to member AB

$$M_{BA} = (2EI/H)(2\theta_B+0-3y_B/H)+0 \qquad (2)$$

From equation (4.3) applied to member BC

$$M_{BC} = (2EI/L)(2\theta_B+\theta_C+0)+0 \qquad (3)$$

From equation (4.4) applied to member BC

$$M_{CB} = 0 = (2EI/L)(2\theta_C+\theta_B+0)+0 \qquad (4)$$

From equilibrium at joint B

$$B) \qquad +M_{BA}+M_{BC} = 0 \qquad (5)$$

At this stage in the calculations there are five equations and six unknowns
and therefore another equation is required. The extra unknown in this problem
is the sway deflection (y_B).
From the equilibrium of member AB (see Fig. 4.6(c))

$$B) \qquad +M_{BA}+M_{AB}+WH/2 = 0 \qquad (6)$$

Note that the equation relates the bending moments to the known load and
excludes the reactions V_A, H_A and V_C because these would introduce
additional unknown quantities.

There are now six equations and six unknowns $(M_{AB}, M_{BA}, M_{BC}, y_B, \theta_B$
and $\theta_C)$.
Eliminating y_B from (1) and (2), eliminating θ_C from (3) and (4), and
combining with (5) and (6)

$$\theta_B = +(WHL/EI)/(12+4L/H) \qquad (7)$$

and from (4)

$$\theta_C = -(WHL/2EI)/(12+4L/H) \tag{8}$$

Substituting (7) and (8) in (3)

$$M_{BC} = -3WH/(12+4L/H) \tag{9}$$

Substituting (7) and (9) in (2) and using the relationship in (5) gives

$$y_B = (WH^3/6EI)(3+4L/H)/(12+4L/H) \tag{10}$$

Substituting (7) and (10) in (1)

$$M_{AB} = -WH(3+2L/H)/(12+4L/H)$$

The bending moment diagram is shown in Fig. 4.6(b).

4.5 Non-rectangular sway frame

Determine the slopes at B and C, the deflection at B, and draw the bending moment diagram for the frame shown in Fig. 4.7(a). This problem involves members not intersecting at right angles.

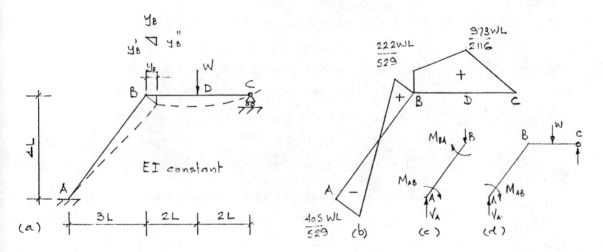

Figure 4.7

Solution From Table 4.1 the fixed end moments are

$$FEM_{BC} = FEM_{CB} = W(4L)/8 = WL/2$$

If the horizontal deflection at B, y_B, is to the right then the deflection perpendicular to AB is

$$y_B' = (5/4)y_B$$

and the deflection perpendicular to BC is

$$y_B'' = (3/4)y_B$$

Form the following equations using the slope deflection sign convention. Where signs are not known then they are assumed to be positive.
From equation (4.3) applied to member AB

$$M_{AB} = (2EI/5L)(0+\theta_B-3y_B'/L)+0$$

Expressing y_B' in terms of y_B

$$M_{AB} = (2EI/5L)(0+\theta_B-3y_B/4L)+0 \qquad (1)$$

Similarly by applying equation (4.4) to member AB

$$M_{BA} = (2EI/5L)(2\theta_B+0-3y_B/4L)+0 \qquad (2)$$

From equation (4.3) applied to member BC

$$M_{BC} = (2EI/4L)(2\theta_B+\theta_C+9y_B/16L)-WL/2 \qquad (3)$$

From equation (4.4) applied to member BC

$$M_{CB} = 0 = (2EI/4L)(2\theta_C+\theta_B+9y_B/16L)+WL/2 \qquad (4)$$

From equilibrium at joint B

$$\text{B)} \quad +M_{BA}+M_{BC} = 0 \qquad (5)$$

The additional sway equation is obtained as follows.
For the equilibrium of member AB (see Fig. 4.7(c))

$$\text{B)} \quad +M_{AB}+M_{BA}+V_A\times 3L = 0 \qquad (i)$$

For the equilibrium of members ABC, (see Fig. 4.7(d))

$$\text{C)} \quad +M_{AB}+V_A\times 7L-W\times 2L = 0 \qquad (ii)$$

combining (i) and (ii) and eliminating V_A

$$+4M_{AB}+7M_{BA}+6WL = 0 \qquad (6)$$

There are six equations and six unknowns (M_{AB}, M_{BA}, M_{BC}, y_B, θ_B and θ_C).
Substituting (1) and (2) in (5) and combining with (3) and (4)

$$\theta_B = +545WL^2/(529EI) \qquad (7)$$

and

$$\theta_C = -2089WL^2/(529EI) \qquad (8)$$

Substituting (7) and (8) in (4)

$$y_B = +1960WL^3/(529EI) \qquad (9)$$

Substituting (7) and (9) in (1)

$$M_{AB} = -405WL/529$$

Substituting (7) and (9) in (2)

$$M_{BA} = -222WL/529$$

The bending moment diagram is shown in Fig. 4.7(b). The bending moment at D is obtained from adjusting the simply supported bending moment by the end bending moments (see Fig. 4.7(b)):

$$M_D = WL/4 + M_B/2$$
$$= WL/4 + 222WL/(529 \times 2) = 973WL/2116$$

4.6 Sway portal frame

Determine the slopes at B and C, the deflection at B, and draw the bending moment diagram for the frame shown in Fig. 4.8(a). As the structure becomes more redundant the number of equations increases and solutions by hand become laborious.

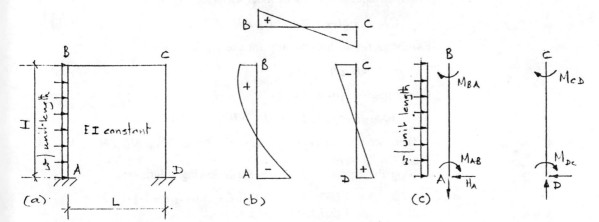

Figure 4.8

Solution Assume sway to the right. Where signs are not known then they are assumed to be positive.

From equation (4.3) applied to member AB

$$M_{AB} = (2EI/H)(0 + \theta_B - 3y_B/H) - wH^2/12 \qquad (1)$$

From equation (4.4) applied to member AB

$$M_{BA} = (2EI/H)(2\theta_B + 0 - 3y_B/H) + wH^2/12 \qquad (2)$$

From equation (4.3) applied to member BC

$$M_{BC} = (2EI/L)(2\theta_B + \theta_C + 0) + 0 \qquad (3)$$

From equation (4.4) applied to member BC

$$M_{CB} = (2EI/L)(2\theta_C + \theta_B + 0) + 0 \qquad (4)$$

From equation (4.3) applied to member CD

$$M_{CD} = (2EI/H)(2\theta_C + 0 - 3y_B/H) + 0 \qquad (5)$$

From equation (4.4) applied to member CD

$$M_{DC} = (2EI/H)(0+\theta_C-3y_B/H)+0 \tag{6}$$

For equilibrium of joint B

B) $M_{BA}+M_{BC} = 0$ $\qquad\qquad$ (7)

For equilibrium of joint C

D) $M_{CB}+M_{CD} = 0$ $\qquad\qquad$ (8)

From the equilibrium of member AB (see Fig. 4.8(c))

B) $+M_{BA}+M_{AB}+WH^2/2+H_A\times H = 0$ $\qquad$ (i)

From the equilibrium of member CD (see Fig. 4.8(c))

C) $+M_{CD}+M_{DC}+H_D\times H = 0$ $\qquad\qquad$ (ii)

Resolving forces horizontally for the whole frame

$\quad\longrightarrow\quad wH-H_A-H_D = 0$ $\qquad\qquad$ (iii)

Combining (i), (ii) and (iii)

$$+M_{AB}+M_{BA}+M_{CD}+M_{DC}+WH^2/2 = 0 \tag{9}$$

There are nine equations and nine unknowns (M_{AB}, M_{BA}, M_{BC}, M_{CB}, M_{CD}, M_{DC}, y_B, θ_B and θ_C).
Solving the equations produces the following solutions:

$$\begin{aligned}
\theta_B &= +[WH^3/(8EI)]\{(2H/L+3)/[6(H/L+2)]\\
&\quad +(2H/L+1)/(6H/L+1)-2/3\}\\
\theta_C &= +WH^3/(8EI)\{-(2H/L+3)/[6(H/L+2)]\\
&\quad +(2H/L+1)/(6H/L+1)\}\\
y_B &= +[WH^4/(16EI)][(2H/L+3)/(6H/L+1)]\\
M_{AB} &= -(WH^2/4)\{(H/L+3)/[6(H/L+2)]+(4H/L+1)/(6H/L+1)\}\\
M_{BA} &= (WH^2/4)\{-(H/L)/[6(H/L+2)]+(2H/L)/(6H/L+1)\}\\
M_{CD} &= -(WH^2/4)\{(H/L)/[6(H/L+2)]+(2H/L)/(6H/L+1)\}\\
M_{DC} &= (WH^2/4)\{-(H/L+3)/[6(H/L+2)]+(4H/L+1)/(6H/L+1)\}
\end{aligned}$$

The bending moment diagram is shown in Fig. 4.8(b).

4.7 Asymmetrical portal frame

Form the slope deflection equations for the asymmetrical portal frame shown in Fig. 4.9.

Solution Assume sway to the right.
From equation (4.3) applied to member AB

$$M_{AB} = (2EI/3L)(0+\theta_B-3y_B/3L)+0 \tag{1}$$

From equation (4.4) applied to member AB

$$M_{BA} = (2EI/3L)(2\theta_B+0-3y_B/3L)+0 \tag{2}$$

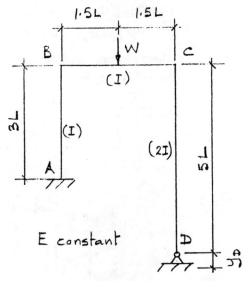

Figure 4.9

From equation (4.3) applied to member BC

$$M_{BC} = (2EI/3L)(2\theta_B + \theta_C - 3y_D/3L) - W(3L)^2/8 \qquad (3)$$

From equation (4.4) applied to member BC

$$M_{CB} = (2EI/3L)(2\theta_C + \theta_B - 3y_D/3L) + W(3L)^2/8 \qquad (4)$$

From equation (4.3) applied to member CD

$$M_{CD} = [2E(2I)/5L](2\theta_C + \theta_D - 3y_B/5L) + 0 \qquad (5)$$

From equation (4.4) applied to member CD

$$M_{DC} = 0 = [2E(2I)/5L](2\theta_D + \theta_C - 3y_B/5L) + 0 \qquad (6)$$

For equilibrium of joint B

$$B) \quad M_{BA} + M_{BC} = 0 \qquad (7)$$

For equilibrium of joint C

$$C) \quad M_{CB} + M_{CD} = 0 \qquad (8)$$

From the equilibrium of member AB

$$B) \quad +M_{BA} + M_{AB} - H_A \times 3L = 0 \qquad (i)$$

From the equilibrium of member CD

$$C) \quad +M_{CD} - H_D \times 5L = 0 \qquad (ii)$$

Resolving horizontally for the whole frame

$$\rightarrow \quad +H_A + H_D = 0 \qquad (iii)$$

Combining (i), (ii) and (iii)

$$(M_{AB}+M_{BA})/(3L)+M_{CD}/(5L) = 0 \qquad (9)$$

There are nine equations and nine unknowns (M_{AB}, M_{BA}, M_{BC}, M_{CB}, M_{CD}, y_B, θ_B, θ_C and θ_D).

4.8 Portal frame with inclined member

Form the slope deflection equations for the frame shown in Fig. 4.10.

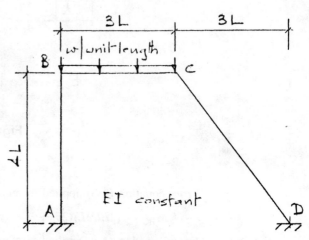

Figure 4.10

Solution Assume sway to the left.
From equation (4.3) applied to member AB

$$M_{AB} = (2EI/4L)(0+\theta_B+3y_B/4L)+0 \qquad (1)$$

From equation (4.4) applied to member AB

$$M_{BA} = (2EI/4L)(2\theta_B+0-3y_B/4L)+0 \qquad (2)$$

From equation (4.3) applied to member BC

$$M_{BC} = (2EI/3L)(2\theta_B+\theta_C+3(3/4)y_B/3L)-w(3L)^2/12 \qquad (3)$$

From equation (4.4) applied to member BC

$$M_{CB} = (2EI/3L)(2\theta_C+\theta_B+3(3/4)y_B/3L)+w(3L)^2/12 \qquad (4)$$

From equation (4.3) applied to member CD

$$M_{CD} = (2EI/5L)(2\theta_C+0-3(5/4)y_B/5L)+0 \qquad (5)$$

From equation (4.4) applied to member CD

$$M_{DC} = (2EI/5L)(0+\theta_C-3/(5/4)y_B/5L)+0 \qquad (6)$$

For equilibrium of joint B

$$\text{B)} \quad M_{BA} + M_{BC} = 0 \tag{7}$$

For equilibrium of joint C

$$\text{C)} \quad M_{CB} + M_{CD} = 0 \tag{8}$$

From the equilibrium of member AB

$$\text{B)} \quad +M_{AB} + M_{BA} - H_A \times 4L = 0 \tag{i}$$

From the equilibrium of member CD

$$\text{C)} \quad +M_{CD} + M_{DC} - H_D \times 4L - V_D \times 3L = 0 \tag{ii}$$

From the equilibrium of members BCD

$$\text{B)} \quad +M_{BC} + M_{DC} - H_D \times 4L - V_D \times 6L + w(3L)^2/2 = 0 \tag{iii}$$

Resolving horizontally for the whole frame

$$\longrightarrow \quad +H_A + H_D = 0 \tag{iv}$$

Combining (i), (ii), (iii), (iv) and (7)

$$+2M_{AB} + 4M_{BA} + 4M_{CD} + 2M_{DC} - 9wL^2 = 0 \tag{9}$$

There are nine equations and nine unknowns (M_{AB}, M_{BA}, M_{BC}, M_{CB}, M_{CD}, M_{DC}, y_B, θ_B and θ_C).

Problems

1 Draw the bending moment diagram and determine the rotation at B for the propped cantilever shown in Fig. 4.11(a).

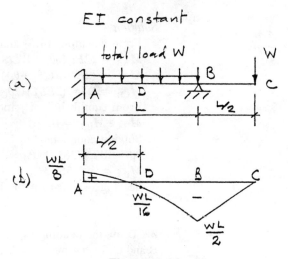

Figure 4.11

Solution
From equations (4.3) and (4.4)
$M_{AB} = (2EI/L)\theta_B - WL/12$, $M_{BA} = (4EI/L)\theta_B + WL/12$,
$\qquad M_{BC} + WL/2 = 0$.
From equilibrium at joint B
$M_{BA} + M_{BC} = 0$.
Solving equations
$\theta_B = 5WL^2/(48EI)$, $M_{AB} = WL/8$, $M_{BA} = WL/2$, $M_D = -WL/16$.
See Fig. 4.11(b) for the bending moment diagram.

2 Draw the bending moment diagram and determine the rotations at B and C for the two span beam shown in Fig. 4.12(a).

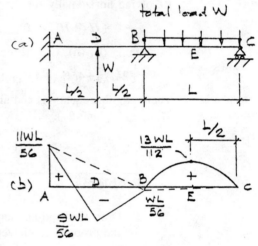

Figure 4.12

Solution
From equations (4.3) and (4.4)
$M_{AB} = (2EI/L)\theta_B + WL/8$, $M_{BA} = (4EI/L)\theta_B - WL/8$,
$M_{BC} = (2EI/L)(2\theta_B + \theta_C) - WL/12$,
$M_{CB} = 0 = (2EI/L)(2\theta_C + \theta_B) + WL/12$.
From equilibrium at joint B
$M_{BA} + M_{BC} = 0$.
Solving equations
$\theta_B = WL^2/(28EI)$, $\theta_C = -13WL^2/(336EI)$, $M_{AB} = 11WL/56$,
$M_{BA} = WL/56$, $M_D = -9WL/56$, $M_E = 13WL/112$.
See Fig. 4.12(b) for the bending moment diagram.

3 Draw the bending moment diagram and determine the rotations at B and C for the two span beam with settlement at support C as shown in Fig. 4.13(a).

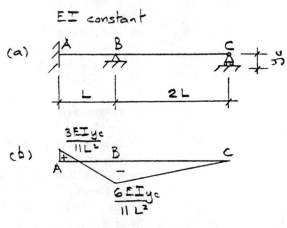

EI constant

(a)

(b)

Figure 4.13

Solution

From equations (4.3) and (4.4)

$M_{AB} = (2EI/L)\theta_B$, $M_{BA} = (4EI/L)\theta_B$,

$M_{BC} = (EI/L)[2\theta_B + \theta_C - 3y_C/(2L)]$,

$M_{CB} = 0 = (EI/L)[2\theta_C + \theta_B - 3y_C/(2L)]$.

From equilibrium at joint B

$M_{BA} + M_{BC} = 0$

Solving equations

$\theta_B = 3y_C/(22L)$, $\theta_C = 15y_C/(22L)$, $M_{AB} = 3EIy_C/(11L^2)$,

$M_{BA} = 6EIy_C/(11L^2)$.

See Fig. 4.13(b) for the bending moment diagram.

4 Draw the bending moment diagram and determine the rotations at B and C for the rectangular sway frame shown in Fig. 4.14(a).

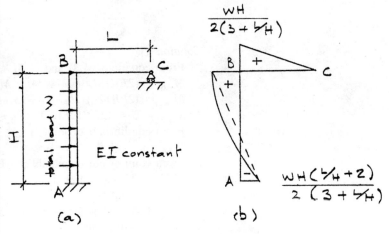

Figure 4.14

Solution

From equations (4.3) and (4.4)

$M_{AB} = (2EI/H)(\theta_B - 3y_B/H) - WH/12$,

$M_{BA} = (2EI/L)(2\theta_B - 3y_B/H) + WH/12$,

$M_{BC} = (2EI/L)(2\theta_B + \theta_C)$,

$M_{CB} = 0 = (2EI/L)(2\theta_C + \theta_B)$.

From equilibrium of joint B

$M_{BA} + M_{BC} = 0$.

From equilibrium of member AB, $M_{AB} + M_{BA} + WH/2 = 0$

Solving equations

$\theta_B = +(WHL/6EI)/(3+L/H)$, $\theta_C = -(WHL/12EI)/(3+L/H)$,

$y_B = (WLH^2/72EI)(13+5L/H)/(3+L/H)$,

$M_{AB} = -(WH/2)(L/H+2)/(3+L/H)$, $M_{BA} = -(WH/2)/(3+L/H)$.

See Fig. 4.14(b) for the bending moment diagram.

5 Draw the bending moment diagram and determine the rotations at B and C for the non-rectangular sway frame shown in Fig. 4.15(a).

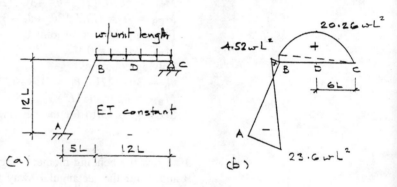

Figure 4.15

Solution

From equations (4.3) and (4.4)

$M_{AB} = (2EI/13L)(\theta_B - y_B/4L)$, $M_{BA} = (2EI/13L)(2\theta_B - y_B/4L)$,

$M_{BC} = (2EI/12L)(2\theta_B + \theta_C + 15y_B/144L) - 12wL^2$,

$M_{CB} = 0 = (2EI/12L)(2\theta_C + \theta_B + 15y_B/144L) + 12wL^2$.

From equilibrium of joint B

$M_{BA} + M_{BC} = 0$.

From equilibrium of members AB and ABC

$12M_{AB} + 17M_{BA} + 360wL^2 = 0$.

Solving equations

$\theta_B = 124.1wL^3/(EI)$, $\theta_C = -155.9wL^3/(EI)$, $M_{AB} = -23.6wL^2$,

$M_{BA} = -4.518wL^2$.

See Fig. 4.15(b) for the bending moment diagram.

6 Form the slope deflection equations for the eccentrically loaded symmetrical portal frame shown in Fig. 4.16.

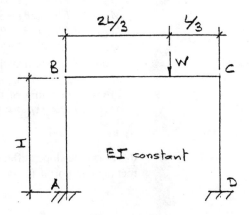

Figure 4.16

Solution

From equations (4.3) and (4.4)

$M_{AB} = (2EI/H)(\theta_B - 3y_B/H)$, $M_{BA} = (2EI/H)(2\theta_B - 3y_B/H)$,

$M_{BC} = (2EI/L)(2\theta_B + \theta_C) - 2WL/27$,

$M_{CB} = (2EI/L)(2\theta_C + \theta_B) + 4WL/27$,

$M_{CD} = (2EI/H)(2\theta_C - 3y_B/H)$, $M_{DC} = (2EI/H)(\theta_C - 3y_B/H)$.

From equilibrium of joints B and C

$M_{BA} + M_{BC} = 0$, $M_{CB} + M_{CD} = 0$.

From equilibrium of the columns

$M_{AB} + M_{BA} + M_{CD} + M_{DC} = 0$.

7 Form the slope deflection equations for the laterally loaded unsymmetrical portal frame shown in Fig. 4.17.

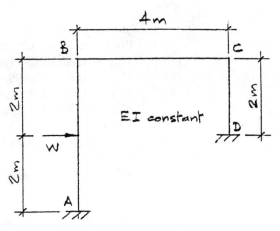

Figure 4.17

Solution

From equations (4.3) and (4.4)

$M_{AB} = (EI/2)(\theta_B - 3y_B/4) - W/2$,

$M_{BA} = (EI/2)(2\theta_B - 3y_B/4 = +W/2)$, $M_{BC} = (EI/2)(2\theta_B + \theta_C)$,

$M_{CB} = (EI/2)(2\theta_C + \theta_B)$, $M_{CD} = (EI)(2\theta_C - 3y_B/4)$,

$M_{DC} = (EI)(\theta_C - 3y_B/4)$.

From equilibrium of joints B and C

$M_{BA} + M_{BC} = 0$, $M_{CB} + M_{CD} = 0$.

From equilibrium of the columns

$M_{AB} + M_{BA} + 2M_{CD} + 2M_{DC} + 2W = 0$

8 Form the slope deflection equations for the eccentrically loaded non-rectangular portal frame shown in Fig. 4.18.

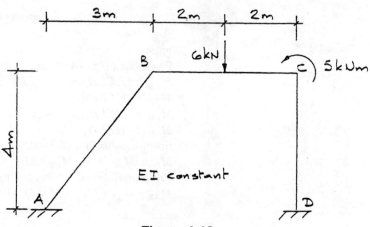

Figure 4.18

Solution

From equations (4.3) and (4.4)

$M_{AB} = (2EI/5)(\theta_B - 3y_B/4)$, $M_{BA} = (2EI/5)(2\theta_B - 3y_B/4)$,

$M_{BC} = (EI/2)(2\theta_B + \theta_C + 9y_B/16) - 3$,

$M_{CB} = (EI/2)(2\theta_C + \theta_B + 9y_B/16) + 3$, $M_{CD} = (EI/2)(2\theta_C - 3y_B/4)$,

$M_{DC} = (EI/2)(\theta_C - 3y_B/4)$.

From equilibrium of joints B and C

$M_{BA} + M_{BC} = 0$, $M_{CB} + M_{CD} + 5$

From equilibrium of the columns

$4M_{AB} + 7M_{BA} + 7M_{CD} + 4M_{DC} + 51 = 0$

9 Form the slope deflection equations for the eccentrically loaded column shown in Fig. 4.19.

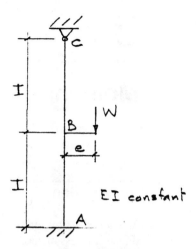

Figure 4.19

Solution

From equations (4.3) and (4.4)

$M_{AB} = (2EI/H)(\theta_B + 3y_B/H)$, $M_{BA} = (2EI/H)(2\theta_B + 3y_B/H)$,

$M_{BC} = (2EI/H)(2\theta_B + \theta_C - 3y_B/H)$,

$M_{CB} = (2EI/H)(2\theta_C + \theta_B - 3y_B/H)$.

From equilibrium of joint B

$M_{BA} + M_{BC} + We = 0$.

From overall equilibrium

$M_{AB} + M_{BA} + M_{BC} = 0$.

5

Moment distribution method

Introduction

Moment distribution is a hand calculation method of structural analysis based on the stiffness method which is applicable to certain types of elastic hyperstatic structures where members are predominantly in bending. The method was very popular prior to the introduction of the computer and is still used for simple structures in design situations. It is generally limited to structures in one plane and assumes that axial and shear forces have little effect on the solution of a structure which is essentially in bending.

The method entails iteratively adjusting the numerical value of moments at the ends of members to maintain equilibrium and is particularly useful for continuous beams. It can also be applied to frames, although the method is laborious where sidesway occurs. The calculations can be terminated when the accuracy is acceptable and it is therefore useful for approximate checks in design calculations. Rotations are not involved directly in the calculations but they can be determined from the bending moment diagrams using the area moment method.

The first step in the method is to split the structure into hyperstatic fixed ended beam cases, e.g. the two span beam shown in Fig. 4.1. The fixed end moments (FEMs) for the load case on each separate beam element are calculated using the values shown in Table 4.1 (see Slope Deflection method).

The separate fixed ended beam elements are reassembled and the end moments at joints, which are free to rotate, are balanced. The out of balance moment is apportioned in relation to the stiffness of the members which intersect the joint.

The stiffness of a member is kEI/L where k is a stiffness factor depending on the end conditions. The stiffnesses of members with different end conditions are given in Table 5.1. Values can be obtained using the area and area-moment method as shown in Examples 3.6 to 3.8.

The proportion of the out of balance fixed end moments which is apportioned to the end of a member is

(out of balance moment) $\times$ (distribution factor)

where distribution factor = (stiffness of a member)/(sum of member stiffnesses of members at a joint).

If a balancing moment is introduced at the end of a member, part of the moment is carried over to the other end of a member (see Example 3.3). The

Table 5.1

Member situation	Stiffness of member AB	Carry over from	
		A to B	A to A'
$+M$ A ⌐)———(⊏ B $+M/2$	EI_{AB}/L	$+M/2$	—
$+M$ A ⌐)———△ B	$\frac{3}{4}EI_{AB}/L$	zero	—
effective fixed end — $+M$ W A ⌐) ↓ B ⌐ W ⊏ $+M/2$	EI_{AB}/L	$+M/2$	—
effective pin — $+M$ W A ⌐) ↓ B ⊏ W	$\frac{3}{4}EI_{AB}/L$	zero	—
effective deflected fixed end — M A M B ⊏	$\frac{1}{2}EI_{AB}/L$	zero	—
effective deflected pin end — W M A M B W	$\frac{3}{2}EI_{AB}/L$	zero	—
effective deflected pin end — M A B M A'	$6EI_{AB}/L$	zero	$-M_{AA'}$

proportion to be carried over and the associated sign is given in Table 5.1. The process of balance and carry over is continued until a satisfactory accuracy is attained.

The sign convention adopted for end moments in the method of moment distribution is clockwise positive, anti-clockwise negative. This convention is *not* the same as the right hand screw rule used for plotting bending moment diagrams (see Volume 1 of *Solving Problems in Structures*).

Worked examples

The first group of examples demonstrates the method of dealing with different types of loading and settlement of supports on two span continuous beams with fixed end and pin supports.

5.1 Fixed ended two span beam

Draw the bending moment and shear force diagrams for the beam shown in Fig. 5.1(a).

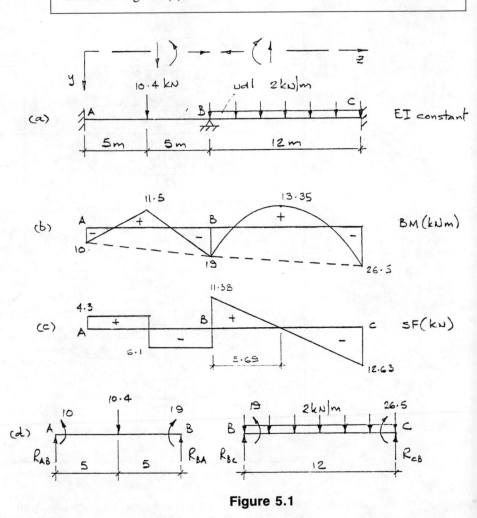

Figure 5.1

Solution

Construct the moment distribution table, as follows.

From Table 4.1 the fixed end moments are

$$\text{FEM}_{AB} = \text{FEM}_{BA} = WL/8 = 10.4 \times 10/8 = 13 \, \text{kNm}$$

Distribution factors		6/11	5/11		
Members	AB	BA	BC	CB	
FEMs	− 13	+13	− 24	+24	kNm
Balance		+ 6	+ 5		
Carry over	+ 3			+ 2.5	
Final moments	− 10	+19	− 19	+26.5	kNm

and
$$FEM_{BC} = FEM_{CB} = WL/12 = (2 \times 12) \times 12/12 = 24 \, kNm$$

The signs inserted in the moment distribution table depend on the direction of the fixed end moments. The direction of the end moment is that required to maintain the ends of the beams in a horizontal position. Positive is clockwise and negative is anticlockwise as shown in Table 4.1.

From Table 5.1 the distribution factors are
$$DF_{BA} = (EI_{BA}/L_{BA})/(EI_{BA}/L_{BA} + EI_{BC}/L_{BC})$$
$$= (EI/10)/(EI/10 + EI/12) = 6/11$$
and
$$DF_{BC} = (EI_{BC}/L_{BC})/(EI_{BC}/L_{BC} + EI_{BA}/L_{BA})$$
$$= (EI/12)/(EI/12 + EI/10) = 5/11$$

Out of balance moment at B $= +13 - 24 = -11 \, kNm$
Balance moment to be distributed at B $= +11 \, kNm$
Moment distributed to end BA $= DF_{BA} \times (+11) = (6/11) \times (+11)$
$$= +6 \, kNm$$
Moment distributed to end BC $= DF_{BC} \times (+11) = (5/11) \times (+11)$
$$= +5 \, kNm$$

From Table 5.1 the carry over factor is $+0.5$. Therefore

Carry over moment to end AB $= +0.5 \times (+6) = +3 \, kNm$
Carry over moment to end CB $= +0.5 \times (+5) = +2.5 \, kNm$.

The final moments in the distribution table are the algebraic sum of the moments at the ends of the members.

The following points should be noted as they form a check on the arithmetic and the method.

(1) After each balance, and at completion of the table, the moments around joint B sum to zero, i.e. $+19 - 19 = 0$.

(2) Fixed end joints A and C are not balanced because they are not free to rotate, i.e. they are fixed ends.

(3) There are no carry over moments from fixed end joints A and C because they are not balanced and therefore the table terminates.

Reactions at the supports are calculated as follows.
(1) Isolate span AB (see Fig. 5.1(d)) and sketch the loading and end moments

obtained in the moment distribution analysis. The reactions are assumed to act upwards.

(2) Take moment of forces about end B and determine the vertical reaction at A:

B⟩ $+10R_{AB}+19-10-10.4\times5 = 0$; hence $R_{AB} = +4.3\,\text{kN}$

The positive sign indicates that the assumed upward direction of the reaction is correct.

(3) Take moment of forces about end A and determine the vertical reaction at B:

A⟩ $-10R_{BA}+19-10+10.4\times5 = 0$; hence $R_{BA} = +6.1\,\text{kN}$

Note that R_{BA} is only part of the reaction at the support B.

(4) Isolate span BC (see Fig. 5.1(d)) and sketch the loading and end moments obtained in the moment distribution analysis.

(5) Take moments of forces about end B and determine the vertical reaction at C:

B⟩ $-12R_{CB}-19+26.5+(2\times12)\times6 = 0$;
hence $R_{CB} = +12.625\,\text{kN}$

(6) Take moment of forces about end C and determine the vertical reaction at B:

C⟩ $+12R_{BC}-19+26.5-(2\times12)\times6 = 0$;
hence $R_{BC} = +11.375\,\text{kN}$

Note that the total reaction at B is $R_{BA}+R_{BC} = +6.1+11.375 = 17.475\,\text{kN}$.

(7) Check by resolving forces vertically for the complete structure. This gives

$\uparrow+4.3-10.4+6.1+11.375-2\times12+12.625 = 0$, which is correct.

Plot the shear force diagram using the sign convention based on a right hand set of axes and which is summarised in the diagram above, Fig. 5.1(a).

For span AB

Resolving forces vertically to the right of A, for the left hand part of the beam

$\downarrow+S_B-R_{AB} = 0$; hence $S = +4.3\,\text{kN}$

Resolving forces vertically slightly to the right of the 10.4 kN load for the left hand part of the beam

$\downarrow+S-R_{AB}+10.4 = 0$; hence $S = -6.1\,\text{kN}$

For span BC

Resolving forces vertically slightly to the right of B for the left hand part of the beam

$\downarrow+S-R_{BC} = 0$; hence $S = +11.375\,\text{kN}$

Resolving forces vertically slightly to the left of B for the left hand part of the beam

$\downarrow+S-R_{BC}+2\times12 = 0$; hence $S = -12.625\,\text{kN}$

There is a uniform variation in shear force between B and C as shown in Fig. 5.1(c).

The position of zero shear force in span BC is

$$z = 11.375 \times 12/(11.375+12.625) = 5.6875 \, \text{m from B}$$

Plot the bending moment diagram using the sign convention based on a right hand set of axes which is summarised in the diagram above, Fig. 5.1(a). Note the direction of the arrows in Fig. 5.1(d) in relation to the positive direction shown in the sign convention above Fig. 5.1(a) and ignore the moment distribution signs. Assume the origin is at the left hand end.

For the left hand part of the beam
For span AB
Taking moments of forces about a point slightly to the right of A

$$\circlearrowright \quad +M_A+10 = 0; \text{ hence } M_A = -10 \, \text{kNm}$$

Take moments of forces about the 10.4 kN load. At this point the shear force is zero and the bending moment is a maximum in span AB, i.e.

$$\circlearrowright \quad +M_{max}+10-R_{AB} \times 5 = 0; \text{ hence } M_{max} = +11.5 \, \text{kNm}$$

Taking moments of forces about a point slightly to the left of B

$$\circlearrowright \quad +M_B+10-R_{AB} \times 10+10.4 \times 5 = 0; \text{ hence } M_B = -19 \, \text{kNm}$$

For span BC
Take moments of forces about a point 5.6875 m from B in span BC. At this point the shear force is zero and the bending moment is a maximum in span BC, i.e.

$$\circlearrowright \quad +M_{max}+19-R_{BC} \times 5.6875+2 \times 5.6875^2/2 = 0;$$
hence
$$\circlearrowright \quad M_{max} = +13.35 \, \text{kNm}$$

Taking moments of forces about a point slightly to the left of C

$$+M_C+19-R_{BC} \times 12+2 \times 12 \times 6 = 0; \text{ hence } M_B = -26.5 \, \text{kNm}$$

5.2 Two span beam with cantilever

Draw the shear force and bending moment diagrams for the beam shown in Fig. 5.2(a).

Solution
Construct the moment distribution table as follows.
For the cantilever the moment at the support C is

$$M_{CD} = -WL = -8 \times 3 = -24 \, \text{kNm}$$

From Table 5.1 the distribution factors are

$$DF_{BA} = (EI_{BA}/L_{BA})/(EI_{BA}/L_{BA}+0.75EI_{BC}/L_{BC})$$
$$= (EI/5)/(EI/5+0.75 \times EI/4) = 16/31$$

and
$$DF_{BC} = (0.75EI_{BC}/L_{BC})/(0.75EI_{BC}/L_{BC}+EI_{BA}/L_{BA})$$
$$= (0.75 \times EI/4)/(0.75 \times EI/4+EI/5) = 15/31$$

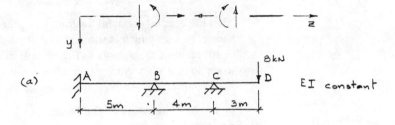

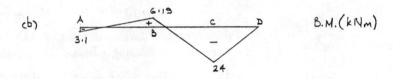

Figure 5.2

Distribution factors			16/31	15/31		
Members	AB		BA	BC	CB	CD
FEMs						−24 kNm
Balance					+24	
Carry over				+12		
Balance			−6.19	−5.81		
Carry over	−3.10					
Final moments	−3.10		−6.19	+6.19	+24	−24 kNm

The procedure is similar to Example 5.1 but with the following variations. Out of balance moment at C = −24 kNm. This moment must remain unaltered at the end of the distribution table and therefore a balancing moment of +24 kNm must be added to CB. There is no carry over from B to C because of the use of $0.75 \times DF_{BC}$.

The reactions at the supports are calculated as follows:

For span AB

B) $+5R_{AB}-3.1-6.19 = 0$; hence $R_{AB} = +1.858$ kN

A) $-5R_{BA}-3.1-6.19 = 0$; hence $R_{BA} = -1.858$ kN
(downwards)

For span BC

$$\text{B)} \qquad -4R_{CB}+6.19+24 = 0; \text{ hence } R_{CB} = +7.5475 \text{ kN}$$
$$\text{C)} \qquad +4R_{BC}+6.19+24 = 0; \text{ hence } R_{BC} = -7.5475 \text{ kN}$$
$$\text{(downwards)}$$

The shear force and bending moment diagrams are constructed as in Example 5.1 and are shown in Figs 5.2(b) and (c).

5.3 Two span beam with support settlement

Draw the bending moment and shear force diagrams for the beam shown in Fig. 5.3(a). This problem illustrates that relative displacements of supports introduce shear force and bending moments into hyperstatic structures.

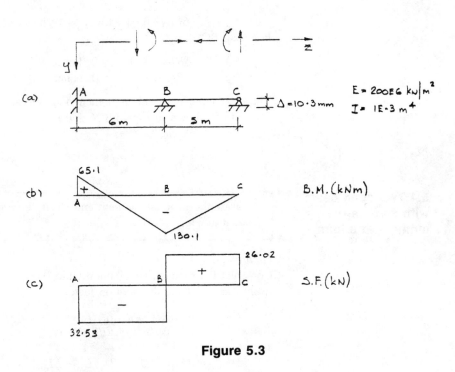

Figure 5.3

Distribution factors			10/19	9/19		
Members	AB	BA	BC	CB		
FEMs			−247.2		kNm	
Balance		+130.1	+117.1			
Carry over	+65.1					
Final moments	+65.1	+130.1	−130.1		kNm	

Solution

Construct the moment distribution table shown above, as follows.
From Table 4.1 the fixed end moment is

$$\text{FEM}_{BC} = 3EI\Delta/L^2 = 3 \times 200\text{E}6 \times 1\text{E-}3 \times 10.3\text{E-}3/5^2 \text{ £ } 247.2 \text{ kNm}$$

From Table 5.1 the distribution factors are

$$DF_{BA} = (EI_{BA}/L_{BA})/(EI_{BA}/L_{BA} + 0.75EI_{BC}/L_{BC})$$
$$= (EI/6)/(EI/6 + 0.75 \times EI/5) = 10/19$$

and

$$DF_{BC} = (0.75EI_{BC}/L_{BC})/(0.75EI_{BC}/L_{BC} + EI_{BA}/L_{BA})$$
$$= (0.75 \times EI/5)/(0.75 \times EI/5 + EI/6) = 9/19$$

The reactions at the supports are determined as follows:
For span AB

$$B\text{)} \quad +6R_{AB} + 65.1 + 130.1 = 0; \text{ hence } R_{AB} = -32.53 \text{ kN}$$
$$\text{(downwards)}$$
$$A\text{)} \quad -6R_{BA} + 65.1 + 130.1 = 0; \text{ hence } R_{BA} = +32.53 \text{ kN}$$

For span BC

$$B\text{)} \quad -5R_{CB} - 130.1 = 0; \text{ hence } R_{CB} = -26.02 \text{ kN (downwards)}$$
$$C\text{)} \quad +5R_{BC} - 130.1 = 0; \text{ hence } R_{BC} = +26.02 \text{ kN}$$

The shear force and bending moment diagrams are constructed as in Example 5.1 and are shown in Figs 5.3(b) and (c).

5.4 Two span beam with an external moment at a joint

Draw the bending moment and shear force diagrams for the beam shown in Fig. 5.4(a).

Solution

Construct the moment distribution table shown as follows.
From Table 5.1 the distribution factors are

$$DF_{BA} = (EI_{BA}/L_{BA})/(EI_{BA}/L_{BA} + EI_{BC}/L_{BC})$$
$$= (EI/6)/(EI/6 + EI/8) = 4/7$$

and

$$DF_{BC} = (EI_{BC}/L_{BC})/(EI_{BC}/L_{BC} + EI_{BA}/L_{BA})$$
$$= (EI/8)/(EI/8 + EI/6) = 3/7$$

It should be noted that in this problem the moment to be distributed at $B = +70$ kNm. It is applied at a joint but is not balanced because it is applied externally.

The reactions at the supports are calculated as follows.
For span AB

$$B\text{)} \quad +6R_{AB} + 40 + 20 = 0; \text{ hence } R_{AB} = -10 \text{ kN (downwards)}$$
$$A\text{)} \quad -6R_{BA} + 40 + 20 = 0; \text{ hence } R_{BA} = +10 \text{ kN}$$

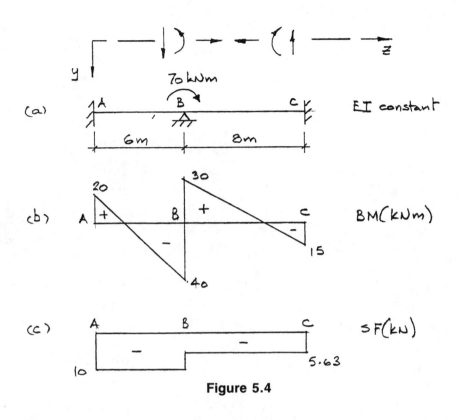

Figure 5.4

Distribution factors		4/7	3/7		
Members	AB	BA	BC	CB	
Distribution		+40	+30		kNm
Carry over	+20			+15	
Final moments	+20	+40	+30	+15	kNm

For span BC

B) $-8R_{CB}+30+15 = 0$; hence $R_{CB} = +5.625$ kN
C) $+8R_{BC}+30+15 = 0$; hence $R_{BC} = -5.625$ kN (downwards)

The shear force and bending moment diagrams are constructed as in Example 5.1 and are shown in Figs 5.4(b) and (c).

5.5 Two span beam with a moment in the span

Draw the bending moment and shear force diagrams for the beam shown in Fig. 5.5(a).

Solution
Construct the moment distribution table shown as follows.

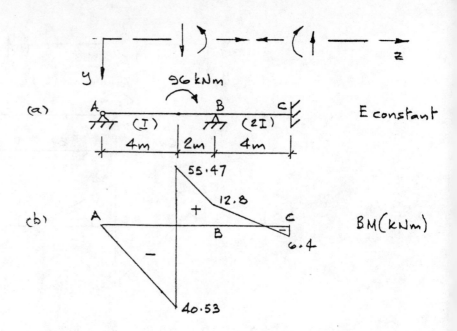

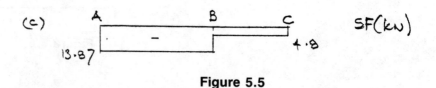

Figure 5.5

Distribution factors		1/5	4/5		
Members	AB	BA	BC	CB	
FEMs		− 16			kNm
Balance		+ 3.2	+12.8		
Carry over				+6.4	
Final moments		− 12.8	+12.8	+6.4	kNm

From Table 4.1 the fixed end moment is

$$\text{FEM}_{BA} = (M/2)[2 - 6b/L + 3(b/L)^2]$$
$$= (96/2)[2 - 6(2/6) + 3(2/6)^2] = 16\,\text{kNm}$$

From Table 5.1 the distribution factors are

$$DF_{BA} = (0.75 \times EI_{BA}/L_{BA})/(0.75 \times EI_{BA}/L_{BA} + EI_{BC}/L_{BC})$$
$$= (0.75 \times EI/6)/(0.75 \times EI/6 + E \times 2I/4) = 1/5$$

and

$$DF_{BC} = (EI_{BC}/L_{BC})/(EI_{BC}/L_{BC} + 0.75 \times EI_{BA}/L_{BA})$$
$$= (E \times 2I/4)(/(E \times 2I/4 + 0.75 \times EI/6)) = 4/5$$

In this problem the moment is not applied at the joint and therefore end moments are calculated and balanced in the normal fashion.

The support reactions are determined as follows.

For span AB

B) $\quad +6R_{AB} - 12.8 + 96 = 0$; hence $R_{AB} = -13.867$ kN (downwards)

A) $\quad -6R_{BA} - 12.8 + 96 = 0$; hence $R_{BA} = +13.867$ kN

For span BC

B) $\quad -4R_{CB} + 12.8 + 6.4 = 0$; hence $R_{CB} = +4.8$ kN

C) $\quad +4R_{BC} + 12.8 + 6.4 = 0$; hence $R_{BC} = -4.8$ kN (downwards)

The shear force and bending moment diagrams are constructed as in Example 5.1 and are shown in Figs 5.5(b) and (c).

The next group of beam problems are more complicated because there are more spans, but the calculations can be reduced if symmetry is present.

5.6 Four span beam with symmetry

> Draw the bending moment and shear force diagrams for the beam shown in Fig. 5.6(a). Calculations for this problem can be reduced if symmetry about joint C is recognised. Treat joint C as a fixed end which does not rotate.

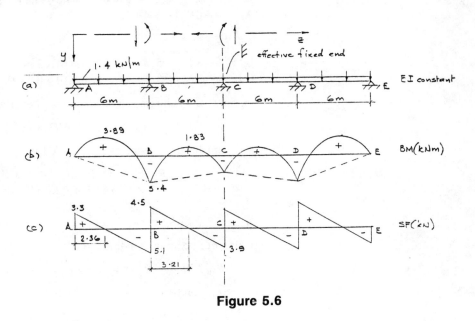

Figure 5.6

Solution

Construct the moment distribution table shown as follows.

Distribution factors		3/7	4/7		
Members	AB	BA	BC	CB	
FEMs		+6.3	−4.2	+4.2	kNm
Balance		−0.9	−1.2		
Carry over				−0.6	
Final moments		+5.4	−5.4	+3.6	kNm

From Table 4.1 the fixed end moments are

$$\text{FEM}_{BA} = WL/8 = (1.4 \times 6) \times 6/8 = 6.3 \text{ kNm}$$

and

$$\text{FEM}_{BC} = FEM_{CB} = WL/12 = (1.4 \times 6) \times 6/12 = 4.2 \text{ kNm}$$

From Table 5.1 the distribution factors are

$$DF_{BA} = (0.75 \times EI_{BA}/L_{BA})/(0.75 \times EI_{BA}/L_{BA} + EI_{BC}/L_{BC})$$
$$= (0.75 \times EI/6)/(0.75 \times EI/6 + EI/6) = 3/7$$

and

$$DF_{BC} = (EI_{BC}/L_{BC})/(EI_{BC}/L_{BC} + 0.75 \times EI_{BA}/L_{BA})$$
$$= (EI/6)/(EI/6 + 0.75 \times EI/6) = 4/7$$

The reactions at the supports are determined as follows.
For span AB

B) $\quad +6R_{AB} + 5.4 - 1.4 \times 6 \times 3 = 0$; hence $R_{AB} = +3.3 \text{ kN}$
A) $\quad -6R_{BA} + 5.4 + 1.4 \times 6 \times 3 = 0$; hence $R_{BA} = +5.1 \text{ kN}$

For span BC

B) $\quad -6R_{CB} - 5.4 + 3.6 + (1.4 \times 6) \times 3 = 0$; hence $R_{CB} = +3.9 \text{ kN}$
C) $\quad +6R_{BC} - 5.4 + 3.6 - (1.4 \times 6) \times 3 = 0$; hence $R_{BC} = +4.5 \text{ kN}$

The position of maximum bending moment in span AB (at zero shear force) is

$$z_1 = R_{AB}L/(R_{AB} + R_{BA}) = 3.3 \times 6/(3.3 + 5.1) = 2.357 \text{ m from A}$$

The maximum bending moment in span AB is

Z1) $\quad +M_{z_1} - R_{AB}z_1 + wz_1^2/2 = 0$
$\quad +M_{z_1} - 3.3 \times 2.357 + 1.4 \times 2.357^2/2 = 0$;

hence

$$M_{z_1} = +3.89 \text{ kNm}$$

The position of maximum bending moment in span BC (at zero shear force) is

$$z_2 = R_{BC}L/(R_{BC} + R_{CB}) = 4.5 \times 6/(4.5 + 3.9) = 3.214 \text{ m from B}$$

The maximum bending moment in span BC is

Z2) $\quad +M_{z_2} + M_{BC} - R_{BC}z_2 + wz_2^2/2 = 0$
$\quad +M_{z_2} + 5.4 - 4.5 \times 3.214 + 1.4 \times 3.214^2/2 = 0$;

hence

$$M_{z2} = +1.83 \text{ kNm}$$

Bending moment and shear force diagrams are shown in Figs 5.6(b) and (c).

5.7 Four span beam with skew-symmetry

Draw the bending moment and shear force diagrams for the beam shown in Fig. 5.7(a). Calculations for this problem can be reduced if skew-symmetry about support C is recognised. Treat support C as a pinned joint.

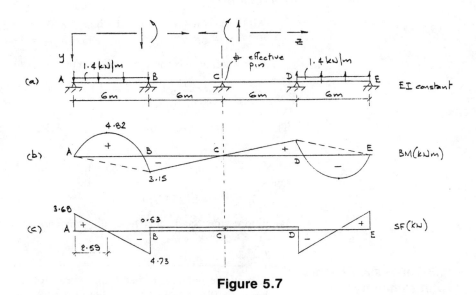

Figure 5.7

Solution

Construct the moment distribution table shown as follows. The table is halved in width because of skew-symmetry.

Distribution factors		1/2	1/2		
Members	AB	BA	BC	CB	
FEMs		+6.3			kNm
Balance		−3.15	−3.15		
Final moments		+3.15	−3.15		kNm

From Table 4.1 the fixed end moment is

$$\text{FEM}_{BA} = WL/8 = (1.4 \times 6) \times 6/8 = 6.3 \text{ kNm}$$

From Table 5.1 the distribution factors are

$$DF_{BA} = (0.75 \times EI_{BA}/L_{BA})/(0.75 \times EI_{BA}/L_{BA} + 0.75 \times EI_{BC}/L_{BC})$$
$$= (0.75 \times EI/6)/(0.75 \times EI/6 + 0.75 \times EI/6) = 1/2$$

and

$$DF_{BC} = (0.75 \times EI_{BC}/L_{BC})/(0.75 \times EI_{BC}/L_{BC} + 0.75 \times EI_{BA}/L_{BA})$$
$$= (0.75 \times EI/6)/(0.75 \times EI/6 + 0.75 \times EI/6) = 1/2$$

The reactions at the supports are determined as follows.
For span AB

B) $+6R_{AB} + 3.15 - (1.4 \times 6) \times 3 = 0$; hence $R_{AB} = +3.675$ kN
A) $-6R_{BA} + 3.15 + (1.4 \times 6) \times 3 = 0$; hence $R_{BA} = +4.725$ kN

For span BC

B) $-6R_{CB} - 3.15 = 0$; hence $R_{CB} = -0.525$ kN (downwards)
C) $+6R_{BC} - 3.15 = 0$; hence $R_{BC} = +0.525$ kN

The position of maximum bending moment in span AB (at zero shear force) is

$$z = R_{AB}L/(R_{AB} + R_{BA}) = 3.675 \times 6/(3.675 + 4.725)$$
$$= 2.589 \text{ m from A}$$

The maximum bending moment in span AB is

Z) $+M_z - R_{AB}z + wz^2/2 = 0$
$+M_z - 3.675 \times 2.589 + 1.4 \times 2.589^2/2 = 0$;

hence

$$M_z = +4.8225 \text{ kNm}$$

Bending moment and shear force diagrams are shown in Figs 5.7(b) and (c).

5.8 Three span beam with symmetry

Draw the bending moment and shear force diagrams for the beam shown in Fig. 5.8(a). Calculations for this problem can be reduced if symmetry about midspan of BC is recognised. Treat mid-span BC as a fixed end which deflects.

Solution
Construct the moment distribution table shown as follows.

Distribution factors			3/5	2/5		
Members	AB		BA	BC	CB	
FEMs				−2.5		kNm
Balance			+1.5	+1.0		
Final moments			+1.5	−1.5		kNm

From Table 4.1 the fixed end moment is

$$FEM_{BC} = WL/12 = (1.2 \times 5) \times 5/12 = 2.5 \text{ kNm}$$

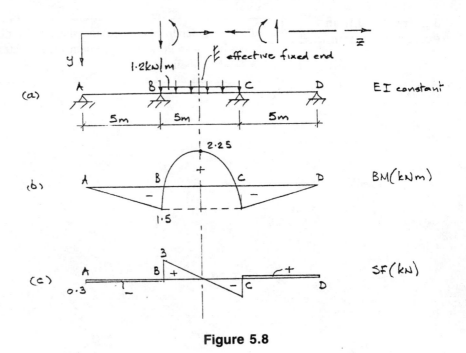

Figure 5.8

From Table 5.1 the distribution factors are

$$DF_{BA} = (0.75 \times EI_{BA}/L_{BA})/(0.75 \times EI_{BA}/L_{BA}+0.5 \times EI_{BC}/L_{BC})$$
$$= (0.75 \times EI/5)/(0.75 \times EI/5+0.5 \times EI/5) = 3/5$$

and

$$DF_{BC} = (0.5 \times EI_{BC}/L_{BC})/(0.5 \times EI_{BC}/L_{BC}+0.75 \times EI_{BA}/L_{BA})$$
$$= (0.5 \times EI/5)/(0.5 \times EI/5+0.75 \times EI/5) = 2/5$$

The bending moment diagram is shown in Fig. 5.8(b). The bending moment at mid-span of BC is given by

$$+WL/8 - (M_B+M_C)/2 = +1.2 \times 5 \times 5/8 - (1.5+1.5)/2$$
$$= +2.25 \text{ kNm}$$

The reactions at the supports are determined as follows.
For span AB

    B⟩ $+5R_{AB}+1.5 = 0$; hence $R_{AB} = -0.3$ kN (downwards)
    A⟩ $-5R_{BA}+1.5 = 0$; hence $R_{BA} = +0.3$ kN

For span BC

    B⟩ $-5R_{CB}-1.5+1.5+1.2 \times 5 \times 2.5 = 0$; hence $R_{CB} = +3.0$ kN

From symmetry

$$R_{BC} = +3.0 \text{ kN}$$

Bending moment and shear force diagrams are shown in Figs 5.8(b) and (c).

5.9 Frame with an eccentric column load

Draw the bending moment and shear force diagrams for the frame shown in Fig. 5.9(a). This problem incorporates an eccentric column load and an inclined member.

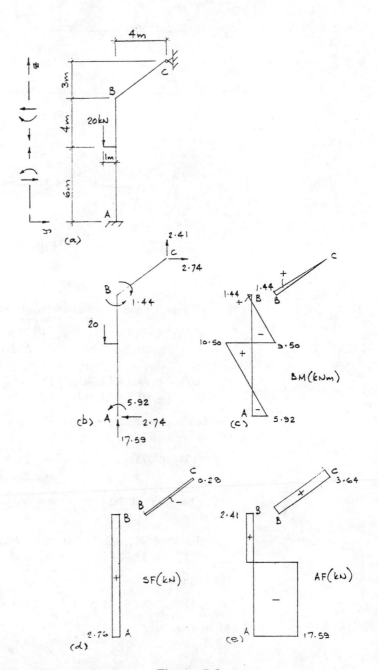

Figure 5.9

Solution

Construct the moment distribution table shown as follows.

Distribution factors		2/5	3/5		
Members	AB	BA	BC	CB	
FEMs	−6.4	−2.4			kNm
Balance		+0.96	+1.44		
Carry over	+0.48				
Final moments	−5.92	−1.44	+1.44		kNm

The eccentric column load is treated as a couple and an axial load applied to member AB. From Table 4.1 the fixed end moments are

$$\text{FEM}_{AB} = +(Mb/L^2)(3a - L) = (-20 \times 4/10^2)(3 \times 6 - 10)$$
$$= -6.4 \text{ kNm}$$

and

$$\text{FEM}_{BA} = (Ma/L^2)(3b - L) = (-20 \times 6/10^2)(3 \times 4 - 10)$$
$$= -2.4 \text{ kNm}$$

From Table 5.1 the distribution factors are

$$DF_{BA} = (EI_{BA}/L_{BA})/(EI_{BA}/L_{BA} + 0.75 \times EI_{BC}/L_{BC})$$
$$= (EI/10)/(EI/10 + 0.75 \times EI/5) = 2/5$$

and

$$DF_{BC} = (0.75 \times EI_{BC}/L_{BC})/(0.75 \times EI_{BC}/L_{BC} + EI_{BA}/L_{BA})$$
$$= (0.75 \times EI/5)/(0.75 \times EI/5 + EI/10) = 3/5$$

The reactions at the supports are determined as follows.
For member AB

$$B) \quad +10H_A - 5.92 - 20 - 1.44 = 0; \text{ hence } H_A = +2.736 \text{ kN}$$

For member BC

$$\rightarrow H_C = H_A = 2.736 \text{ kN}$$
$$B) \quad -4V_C + 1.44 + H_C \times 3 = 0; \text{ hence } V_C = +2.412 \text{ kN}$$

For the complete frame

$$\uparrow V_A = 20 - V_C = 20 - 2.412 = 17.59 \text{ kN}$$

Bending moment and shear force diagrams are shown in Figs 5.9(b) and (c).

The next group of problems introduce further complications associated with vertical members and lateral sway of the structure. As previously if symmetry is present and recognised, then the calculations can be reduced.

5.10 Symmetrical portal frame

Draw the bending moment, shear force and axial force diagrams for the frame shown in Fig. 5.10(a). The frame and loading is symmetrical about the span BC. In this problem three members intersect at joints B and C.

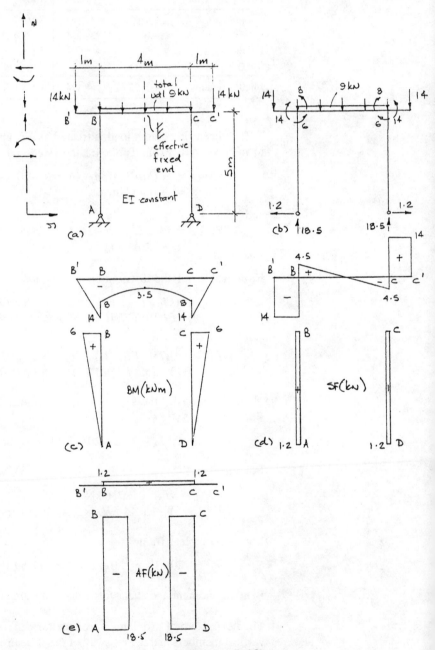

Figure 5.10

Solution
Construct half a moment distribution table as follows.

Distribution factors			6/11		5/11	
Members	AB		BA	BB′	BC	
FEMs				+14	−3	kNm
Balance			−6		−5	
Final moments			−6	+14	−8	kNm

From Table 4.1 the fixed end moments are

$$\text{FEM}_{BC} = WL/12 = 9 \times 4/12 = 3 \text{ kNm}$$

and

$$\text{FEM}_{BB} = WL = 14 \times 1 = 14 \text{ kNm}$$

From Table 5.1 the distribution factors are

$$DF_{BA} = (0.75 \times EI_{BA}/L_{BA})/(0.75 \times EI_{BA}/L_{BA} + 0.5 \times EI_{BC}/L_{BC})$$
$$= (0.75 \times EI/5)/(0.75 \times EI/5 + 0.5 \times EI/4) = 6/11$$

and

$$DF_{BC} = (0.5 \times EI_{BC}/L_{BC})/(0.5 \times EI_{BC}/L_{BC} + 0.75 \times EI_{BA}/L_{BA})$$
$$= (0.5 \times EI/4)/(0.5 \times EI/4 + 0.75 \times EI/5) = 5/11$$

The external forces and the moments at the ends of members are shown in Fig. 5.10(b). The following information is also required.
For member AB

$$\text{B)} \quad +5H_{AB} - 6 = 0; \text{ hence } H_{AB} = +1.2 \text{ kN (outwards)}$$

From symmetry the vertical reactions at the supports are

$$V_A = V_D = (2 \times 14 + 9)/2 = 18.5 \text{ kN}$$

The bending moment diagram is shown in Fig. 5.10(c).
The maximum bending moment at the mid-span of AB is

$$M_{max} = WL/8 - (M_B + M_C)/2 = +9 \times 4/8 - (8+8)/2 = -3.5 \text{ kNm}$$

The shear force diagram shown in Fig. 5.10(d) is drawn using information in Fig. 5.10(b) and the following.
For member BB′

$$\downarrow +S_{BB′} + 14 = 0; \text{ hence } S_{BB′} = -14 \text{ kN}$$

For member BC

$$\downarrow +S_{BC} + 14 - 18.5 = 0; \text{ hence } S_{BC} = +4.5 \text{ kN}$$

The change from B to C is linear.
The axial force diagram is drawn from the forces shown in Fig. 5.10(e).

Draw the bending moment, shear force and axial force diagrams for the frame shown in Fig. 5.11(a). The frame is symmetrical about the span BC but the loading produces sway.

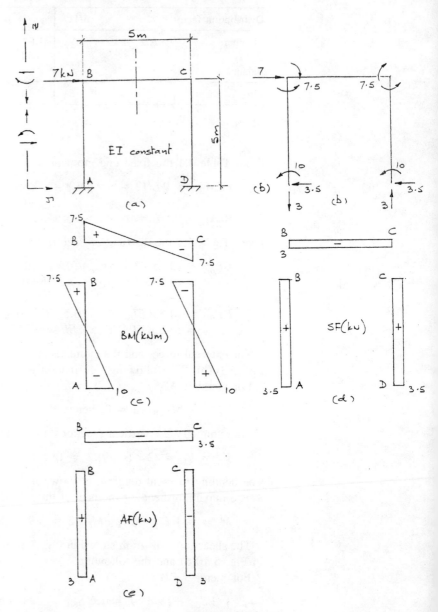

Figure 5.11

Solution
Construct half a moment distribution table as follows.

Distribution factors		1/7	6/7	
Members	AB	BA	BC	
FEMs	− 8.75	−8.75		kNm
Balance		+1.25	+7.5	
Carry over	− 1.25			
Final moments	− 10.00	−7.5	+7.5	kNm

From symmetry the horizontal forces at A and D are equal to $7/2 = 3.5$ kN, and

$$\text{FEM}_{AB} = \text{FEM}_{BA} = \text{FEM}_{CD} = \text{FEM}_{DC} = WH/4 = 7 \times 5/4 = 8.75 \text{ kNm}$$

From Table 5.1 the distribution factors are

$$DF_{BA} = (EI_{BA}/L_{BA})/(EI_{BA}/L_{BA} + 6 \times EI_{BC}/L_{BC})$$
$$= (EI/5)/(EI/5 + 6 \times EI/5) = 1/7$$

and

$$DF_{BC} = (6 \times EI_{BC}/L_{BC})/(6 \times EI_{BC}/L_{BC} + EI_{BA}/L_{BA})$$
$$= (6 \times EI/5)/(6 \times EI/5 + EI/5) = 6/7$$

From Table 5.1 the carry over factor is -1.

Carry over moment to end AB $= -1 \times 1.25 = -1.25$ kNm

There is no carry over moment to end CB because of symmetry.

The forces and end moments shown in Fig. 5.11(b) are obtained from the distribution table and the following calculations.
For member AB

B) $\quad +5H_A - 10 - 7.5 = 0$; hence $H_A = +3.5$ kN (outwards)

For members BCD

B) $\quad -5R_D + 7.5 - 10 + 3.5 \times 5 = 0$; hence $R_D = +3$ kN (upwards).

The bending moment diagram is shown in Fig. 5.11(c).

The shear force diagram shown in Fig. 5.11(d) is constructed from the following information.
For member AB

→ $+S_{AB} - H_A = 0$; hence $S_{AB} = +3.5$ kN

For member BC

↓ $+ S_{BC} + R_A = 0$; hence $S_{BC} = -3$ kN

The axial force diagram (Fig. 5.11(e)) is drawn from the forces shown in Fig. 5.11(b) and the following.
For member AB

↑ $+T_{AB} - R_A = 0$; hence $T_{AB} = +3$ kN

For member BC

$$\rightarrow +T_{BC}+H_A+7 = 0; \text{ hence } T_{BC} = -3.5 \text{ kN}$$

For member CD

$$\uparrow +T_{BC}+R_D = 0; \text{ hence } T_{BC} = -3 \text{ kN}$$

5.12 Symmetrical portal frame with asymmetrical loading

Draw the bending moment, shear force and axial force diagrams for the frame shown in Fig. 5.12(a). The frame is symmetrical about the span BC but the loading is asymmetrical.

The frame can be solved using a full distribution table but the table will be extensive because there will be numerous carry overs. Alternatively the asymmetrical loading is equivalent to the three cases shown in Fig. 5.12(a). This means that three distribution tables are required but symmetry and skew-symmetry reduce the carry overs.

Solution — Case (i)
Construct half a moment distribution table as follows.

Distribution factors		2/3	1/3	
Members	AB	BA	BC	
FEMs	−2.1	+2.1		kNm
Balance		−1.4	−0.7	
Carry over	−0.7			
Final moments	−2.8	+0.7	−0.7	kNm

From Table 4.1 the fixed end moments are

$$\text{FEM}_{AB}=\text{FEM}_{BA}=\text{FEM}_{CD}=\text{FEM}_{DC}=WL/12=4.2\times6/12=2.1 \text{ kNm}$$

From Table 5.1 the distribution factors are

$$DF_{BA} = (EI_{BA}/L_{BA})/(EI_{BA}/L_{BA}+0.5\times EI_{BC}/L_{BC})$$
$$= (EI/6)/(EI/6+0.5\times EI/6) = 2/3$$

and

$$DF_{BC} = (0.5\times EI_{BC}/L_{BC})/(0.5\times EI_{BC}/L_{BC}+EI_{BA}/L_{BA})$$
$$= (0.5\times EI/6)/(0.5\times EI/6+0.75\times EI/6) = 1/3$$

Solution — Case (ii)
Construct half a moment distribution table as follows.
From Table 4.1 the fixed end moments are

$$\text{FEM}_{AB}=\text{FEM}_{BA}=\text{FEM}_{CD}=\text{FEM}_{DC}=WL/12=4.2\times6/12=2.1 \text{ kNm}$$

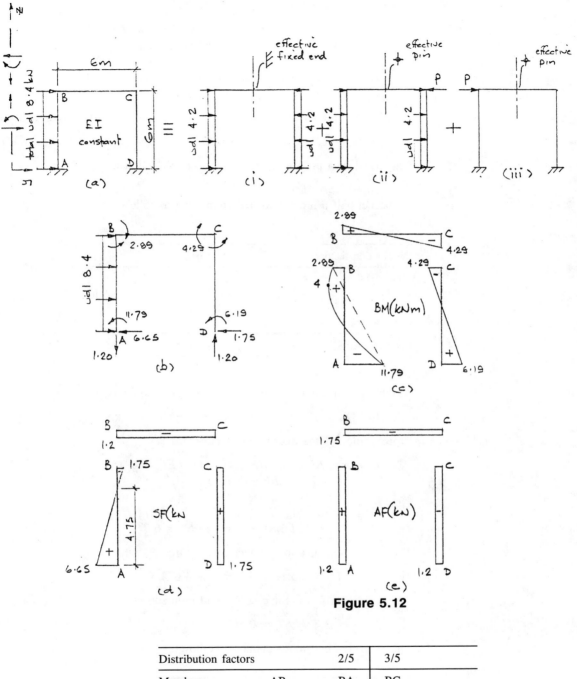

Figure 5.12

Distribution factors		2/5	3/5	
Members	AB	BA	BC	
FEMs	−2.1	+2.1		kNm
Balance		−0.84	−1.26	
Carry over	−0.42			
Final moments	−2.52	+1.26	−1.26	kNm

From Table 5.1 the distribution factors are

$$DF_{BA} = (EI_{BA}/L_{BA})/(EI_{BA}/L_{BA}+0.75 \times EI_{BC}/L_{BC})$$
$$= (EI/6)/(EI/6+0.75 \times EI/6) = 2/5$$

and

$$DF_{BC} = (0.75 \times EI_{BC}/L_{BC})/(0.75 \times EI_{BC}/L_{BC}+EI_{BA}/L_{BA})$$
$$= (0.75 \times EI/6)/(0.75 \times EI/6+EI/6) = 3/5$$

Considering the whole frame the prop force P for case (ii) is obtained from

$$\text{A)} \quad -6P+2 \times (-2.52+1.26)+8.4 \times 3 = 0; \text{ hence } P = 3.78 \text{ kN}$$

The effect of a prop force in the opposite direction is as follows.

Solution — Case (iii)
Construct half a moment distribution table as follows.

Distribution factors		1/7	6/7	
Members	AB	BA	BC	
FEMs	−7	−7		kNm
Balance		+1	+6	
Carry over	−1			
Final moments	−8	−6	+6	kNm

Assume a convenient value for the fixed end moments, i.e.

$$\text{FEM}_{AB} = \text{FEM}_{BA} = \text{FEM}_{CD} = \text{FEM}_{DC} = 7 \text{ kNm}$$

From Table 5.1 the distribution factors are

$$DF_{BA} = (EI_{BA}/L_{BA})/(EI_{BA}/L_{BA}+6 \times EI_{BC}/L_{BC})$$
$$= (EI/6)/(EI/6+6 \times EI/6) = 1/7$$

and

$$DF_{BC} = (6 \times EI_{BC}/L_{BC})/(6 \times EI_{BC}/L_{BC}+EI_{BA}/L_{BA})$$
$$= (6 \times EI/6)/(6 \times EI/6+EI/6) = 6/7$$

These end moments are produced by a force P' where

$$\text{A)} \quad P' = 2 \times (-8-6)/6 = -4.67 \text{ kN}$$

The correction factor for case (ii) end moments is

$$P/P' = 3.78/4.67 = 0.809$$

The corrected end moments are as follows.

	AB	BA	BC
End moments	−6.47	−4.85	+4.85 kNm

Summation of the end moments for cases (i), (ii) and (iii) is shown below.

	AB	BA	BC	CB	CD	DC
End moms (i)	-2.8	$+0.7$	-0.7	$+0.7$	-0.7	$+2.8$
End moms (ii)	-2.52	$+1.26$	-1.26	-1.26	$+1.26$	-2.52
End moms (iii)	-6.47	-4.85	$+4.85$	$+4.85$	-4.85	-6.47
Total	-11.79	-2.89	$+2.89$	$+4.29$	-4.29	-6.19

The forces and end moments shown in Fig. 5.12(b) are obtained from the summation table of end moments and the following.

For member AB

$$\text{B} \rlap{)} \qquad +6H_{AB}-11.79-2.89-8.4\times3 = 0;$$

hence

$$H_A = +6.65\,\text{kN (outwards)}$$

For member CD

$$\text{B} \rlap{)} \qquad +6H_D-4.29-6.19 = 0; \text{ hence } H_D = +1.75\,\text{kN (inwards)}$$

For members BCD

$$\text{B} \rlap{)} \qquad -6V_D+1.75\times6+2.89-6.19 = 0;$$

hence

$$V_D = 1.20\,\text{kN (upwards)}.$$

The bending moment diagram is shown in Fig. 5.12(c).

The shear force diagram shown in Fig. 5.12(d) is constructed from the following information.

For member AB

$$\rightarrow \ +S_A-H_A = 0; \text{ hence } S_{AB} = +6.65\,\text{kN}$$
$$\rightarrow \ +S_B-H_A+8.4 = 0; \text{ hence } S_B = -1.75\,\text{kN}$$

There is a linear variation between A and B.

For member BC

$$\text{↓} \ +S_{BC}+V_A = 0; \text{ hence } S_{BC} = -1.2\,\text{kN}$$

For member CD

$$\leftarrow \ +S_{CD}+H_A-8.4 = 0; \text{ hence } S_{CD} = +1.75\,\text{kN}$$

The position of maximum bending moment (zero shear force) in AB is

$$z = LS_A/(S_A+S_B) = 6\times6.65/(6.65+1.75) = 4.75\,\text{m from A}$$

The maximum bending moment at this point is

$$\text{Z} \rlap{)} \qquad +M_Z+M_A-H_Az+wz^2/2 = 0$$
$$+M_Z+11.79-6.65\times4.75+(8.4/6)\times4.75^2/2 = 0$$

Hence $M_Z = +4.00\,\text{kNm}$

The axial force diagram (Fig. 5.12(e)) is drawn from the forces shown in Fig. 5.12(b) and the following.

For member AB

$$\uparrow +T_{AB}-V_A = 0; \text{ hence } T_{AB} = +1.20 \text{ kN}$$

For member BC

$$\rightarrow +T_{BC}+8.4-6.65 = 0; \text{ hence } T_{BC} = -1.75 \text{ kN}$$

For member CD

$$\downarrow +T_{CD}+V_A = 0; \text{ hence } T_{CD} = -1.2 \text{ kN}$$

Alternatively this problem can be solved by ignoring the symmetry and completing two distribution tables for the complete frame propped and unpropped.

5.13 Asymmetrical portal frame

Draw the bending moment, shear force and axial force diagrams for the frame shown in Fig. 5.13(a). This frame is asymmetrical and a full distribution table is required.

Solution
Construct the moment distribution table for the whole frame as follows.

Distribution factors		1/2	1/2	2/3	1/3	
Members	AB	BA	BC	CB	CD	DC
FEMs			− 12	+12		
Balance		+6	+ 6	− 8	−4	
Carry over	+3		− 4	+ 3		
Balance		+2	+ 2	− 2	−1	
Carry over	+1		− 1	+ 1		
Balance		+0.5	+ 0.5	− 0.7	−0.3	
Final moments	+4	+8.5	− 8.5	+ 5.3	−5.3	

From Table 4.1 the fixed end moments are

$$\text{FEM}_{BC} = \text{FEM}_{CB} = WL/8 = 12 \times 8/8 = 12 \text{ kNm}$$

From Table 5.1 the distribution factors are

$$DF_{BA} = (EI_{BA}/L_{BA})/(EI_{BA}/L_{BA}+EI_{BC}/L_{BC})$$
$$= (EI/8)/(EI/8+EI/8) = 1/2$$

and

$$DF_{CD} = (0.75 \times EI_{BC}/L_{BC})/(0.75 \times EI_{BC}/L_{BC}+EI_{BA}/L_{BA})$$
$$= (0.75 \times EI/12)/(0.75 \times EI/12+EI/8) = 1/3$$

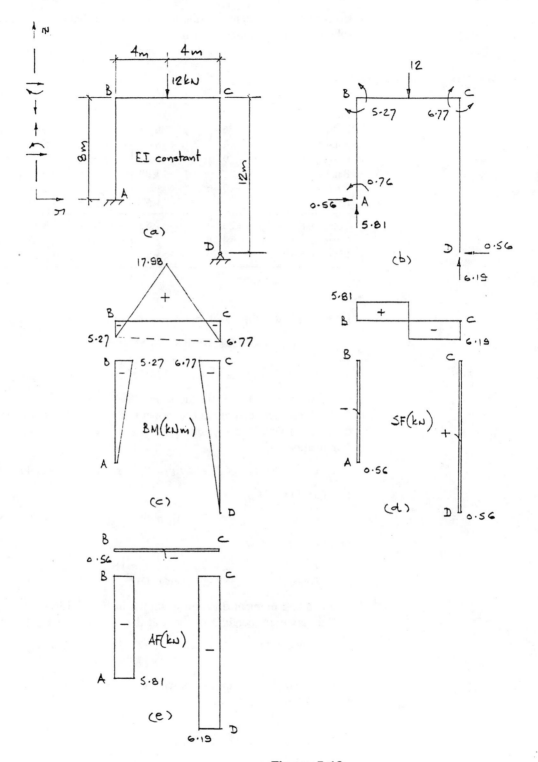

Figure 5.13

Construct the moment distribution table shown below for the whole frame subject to the action of prop forces.

Distribution factors			1/2	1/2	2/3	1/3	
Members	AB	BA	BC	CB	CD		DC
FEMs	− 36	− 36			− 8		
Balance		+18	+18	+5.33	+2.67		
Carry over	+ 9		+ 2.66	+9			
Balance		− 1.33	− 1.33	−6	−3		
Carry over	+ 0.66		− 3	−0.66			
Balance		+ 1.5	+ 1.5	+0.44	+0.22		
Total moments	+26.34	− 17.83	+17.83	+8.11	−8.11		
Corrected moments	− 4.76	− 3.23	+ 3.23	+1.47	−1.47		
Final moments	− 0.76	+ 5.27	− 5.27	+6.77	−6.77		

From Table 4.1 the fixed end moments are

$$\text{FEM}_{AB} = 6EI\Delta/L^2 = 6 \times 384/8^2 = 36 \text{ kNm}$$

and

$$\text{FEM}_{CD} = 3EI\Delta/L^2 = 3 \times 384/12^2 = 8 \text{ kNm}$$

The distribution factors are the same.

The forces and end moments shown in Fig. 5.13(b) are obtained from the distribution table and the following.

For member AB

B⤸ $+8H_A+5.27-0.76 = 0$; hence $H_A = -564$ kN (inwards)

For member CD

C⤸ $+12H_D-6.77 = 0$; hence $H_D = +0.564$ kN (inwards)

For members BCD

B⤸ $-8V_D+12 \times 4 - 5.27 + 0.564 \times 12 = 0$;
hence $V_D = 6.187$ kN (upwards)

The bending moment diagram is shown in Fig. 5.13(c).

The maximum bending moment at the mid-span of BC is

$$+WL/4 - (M_{BC}+M_{DC})/2 = +12 \times 8/4 - (5.27+6.77)/2$$
$$= +17.98 \text{ kNm}$$

The shear force diagram shown in Fig. 5.13(d) is constructed from the following information.

For member AB

→ $+S_{AB}+H_A = 0$; hence $S_{AB} = -0.56$ kN

For member BC

$$\downarrow +S_B - V_A = 0; \text{ hence } S_B = +5.81 \text{ kN}$$
$$\downarrow +S_C - V_A + 12 = 0; \text{ hence } S_B = -6.19 \text{ kN}$$

For member CD

$$\leftarrow +S_{CD} - H_A = 0; \text{ hence } S_{CD} = +0.56 \text{ kN}$$

The axial force diagram (Fig. 5.13(e)) is drawn from the forces shown in Fig. 5.13(b) and the following.
For member AB

$$\uparrow +T_{AB} + V_A = 0; \text{ hence } T_{AB} = -5.81 \text{ kN}$$

For member BC

$$\rightarrow +T_{BC} + H_A = 0; \text{ hence } T_{BC} = -0.56 \text{ kN}$$

For member CD

$$\downarrow +T_{BC} - V_A + 12 = 0; \text{ hence } T_{BC} = -6.19 \text{ kN}$$

5.14 Symmetrical pitched roof portal frame

Draw the bending moment, shear force and axial force diagrams for the frame shown in Fig. 5.14(a). The frame and loading is symmetrical about the apex C and therefore joint C is treated as a fixed end.

There are two stages in the analysis (a) propped at B and D to prevent column sway, and (b) effect of removal of props at B and D.

Solution — Stage (a)
Construct half a moment distribution table as follows, assuming that C is fixed.

Distribution factors			6/2	1/2	
Members	AB	BA	BC	CB	
FEMs			−4	+4	kNm
Balance		+2	+2		
Carry over	+1			+1	
Final moments	+1	+2	−2	+5	kNm

From Table 4.1 the fixed end moment is

$$\text{FEM}_{BC} = (W\cos\theta)(L/\cos\theta)/12 = WL/12 = 4 \times 12/12 = 4 \text{ kNm}$$

From Table 5.1 the distribution factors are

$$DF_{BA} = (EI_{BA}/L_{BA})/(EI_{BA}/L_{BA} + EI_{BC}/L_{BC})$$
$$= (EI/15)/(EI/15 + EI/15) = 1/2$$

and

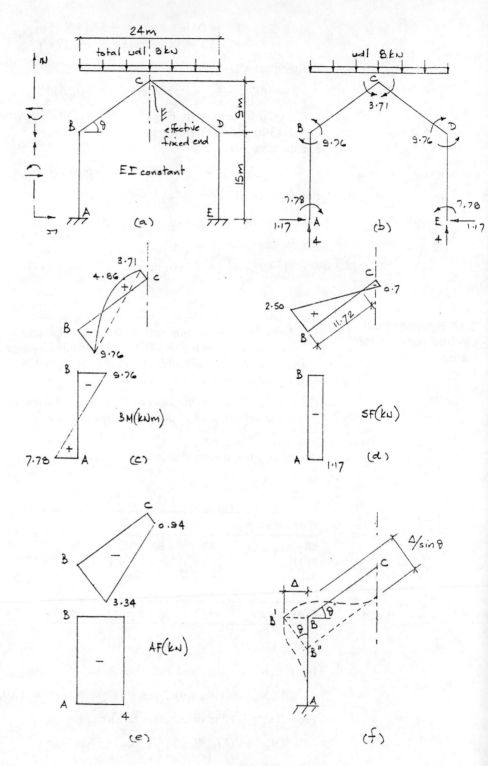

Figure 5.14

$$DF_{BC} = (EI_{BC}/L_{BC})/(EI_{BC}/L_{BC}+EI_{BA}/L_{BA})$$
$$= (EI/15)/(EI/15+EI/15) = 1/2$$

The horizontal prop force P present at B and D for case (a) end moments is obtained as follows:

From symmetry and resolving forces vertically

$$V_A = V_D = 4\,kN$$

For member AB

B$\big)$ $-15H_A+2+1 = 0$; hence $H_A = 0.2\,kN$ (inwards)

For members ABC

C$\big)$ $-9P+5+1-0.2\times24+4\times12-4\times6 = 0$;
hence $P = 2.8\,kN$ (inwards)

Solution — Stage (b)

To determine the effect of an arbitrary prop force P', applied at B and D, construct half a moment distribution table as follows.

Distribution factors		1/2	1/2		
Members	AB	BA	BC	CB	
FEMs	+30	+30	−50	−50	kNm
Balance		+10	+10		
Carry over	+ 5			+ 5	
Final moments	+35	+40	−40	−45	kNm

From Table 4.1 the fixed end moment (see Fig. 5.14(f)) is

$$FEM_{AB}: FEM_{BC} = 6EI\Delta/15^2:(6EI\Delta/\sin\theta)/15^2 = 3:5$$

The distribution factors are as case (a).

The horizontal prop force P' present at B and D for case (b) moments is obtained as follows.

From symmetry and resolving forces vertically

$$V_A = V_D = 0$$

For member AB

B$\big)$ $-15H_A+40+35 = 0$; hence $H_A = 5\,kN$ (inwards)

For member ABC

C$\big)$ $-9P'-45+35-5\times24 = 0$; hence $P' = 14.44\,kN$ (outwards)

The correction factor for sway moments is

$$P/P' = 2.8/14.44 = 0.194$$

Summation of moments from cases (a) and (b) is shown below.

Members	AB	BA	BC	CB	
Case (a)	+1	+2	−2	+5	kNm
Case (b)					
corrected	+6.78	+7.76	−7.76	−8.71	
Final moments	+7.78	+9.76	−9.76	−3.71	kNm

From symmetry the vertical reactions at the supports are

$$\uparrow V_A = V_D = 4\,\text{kN}$$

For member AB

$$\text{B)} \quad -15H_A + 9.76 + 7.78 = 0; \text{ hence } H_A = +1.169\,\text{kN (inwards)}$$

Forces and moments are shown in Fig. 5.14(b) and the bending moment diagram is shown in Fig. 5.14(c).

The shear force diagram shown in Fig. 5.14(d) is constructed from the following information.
For member AB

$$\rightarrow \quad +S_{AB} + H_A = 0; \text{ hence } S_{AB} = -1.169\,\text{kN}$$

for member BC

$$\searrow \quad +S_B + 1.169 \times 0.6 - 4 \times 0.8 = 0; \text{ hence } S_B = +2.498\,\text{kN}$$
$$\searrow \quad +S_C + 1.169 \times 0.6 - 4 \times 0.8 + 4 \times 0.8 = 0; \text{ hence } S_C = -0.702\,\text{kN}$$

The change in shear force from B to C is linear.

The position of maximum bending moment (zero shear) in BC occurs at

$$z(\text{slope}) = LS_B/(S_B + S_C) = 15 \times 2.498/(2.498 + 0.702) = 11.72\,\text{m from B}$$
$$z(\text{horizontal}) = z(\text{slope})\cos\theta = 11.72 \times \cos\theta = 11.72 \times 0.8 = 9.38\,\text{m}$$
$$\text{from B}$$
$$z(\text{vertical}) = z(\text{slope})\sin\theta = 11.72 \times \sin\theta = 11.72 \times 0.6 = 7.03\,\text{m from B}$$

The maximum bending moment in member BC is

$$\text{Z)} \quad +M_{max} + (8/24) \times 9.38^3/2 - 4 \times 9.38 + 1.17 \times (15 + 7.03) - 7.78 = 0$$
$$M_{max} = +4.86\,\text{kNm}$$

The axial force diagram (see Fig. 5.14(e)) is drawn using the forces shown in Fig. 5.14(b).
For member AB

$$\uparrow \quad +T_{AB} + V_A = 0; \text{ hence } T_{AB} = -4\,\text{kN}$$

For member BC

$$\nearrow \quad +T_B + 1.17 \times 0.8 + 4 \times 0.6 = 0; \text{ hence } T_B = -3.336\,\text{kN}$$
$$\nearrow \quad +T_C + 1.17 \times 0.8 - 4 \times 0.6 + 4 \times 0.6 = 0; \text{ hence } T_C = -0.936\,\text{kN}$$

The change in axial force from B to C is linear.

Problems

Draw the bending moment, shear force and axial force diagrams for the structures shown.

1 See Fig. 5.15.

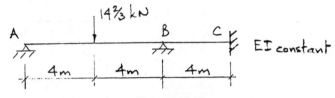

Figure 5.15

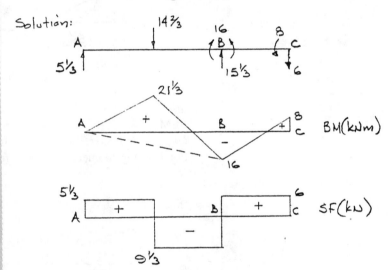

Figure 5.16

Solution

$\text{FEM}_{BA} = +3WL/16 = +22 \text{ kNm}$

$DF_{BA} = (3/4)(EI/8)/[(3/4)(EI/8)+EI/4] = 3/11.$

For bending moment and shear force diagrams see Fig. 5.16.

2 See Fig. 5.17.

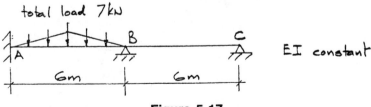

Figure 5.17

Solution:

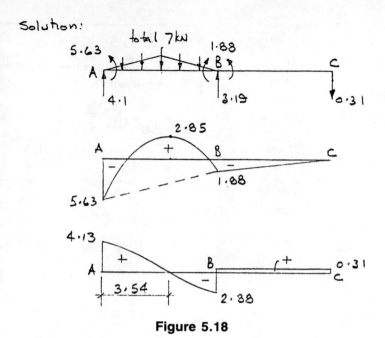

Figure 5.18

Solution

$FEM_{AB} = -5WL/48 = -4.375\,kNm$, $FEM_{BA} = +4.375\,kNm$.
$DF_{BA} = (EI/6)/[(EI/6)+(3/4)(EI/6)] = 4/7$.

For bending moment and shear force diagrams see Fig. 5.18.

3 See Fig. 5.19.

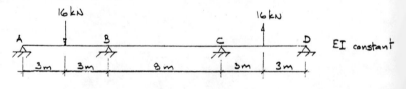

Figure 5.19

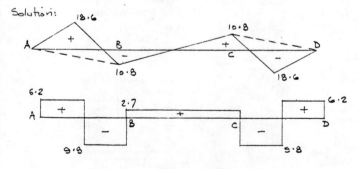

Figure 5.20

Skew-symmetry about centre of span BC.

$FEM_{BA} = +3WL/16 = +18\,kNm.$

$DF_{BA} = (3/4)(EI/6)/[(3/4)(EI/6)+(3/2)(EI/8)] = 2/5.$

For bending moment and shear force diagrams see Fig. 5.20.

4 See Fig. 5.21.

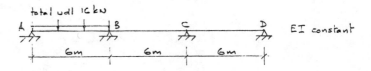

Figure 5.21

Solution:

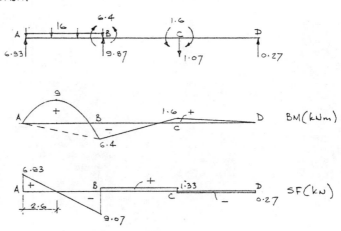

Figure 5.22

Solution

Symmetry about centre of span BC.

$FEM_{BA} = +WL/8 = +6\,kNm.$

$DF_{BA} = (3/4)(EI/6)/[(3/4)(EI/6)+(1/2)(EI/6)] = 3/5.$

Skew-symmetry about centre of span BC.

$FEM_{BA} = +WL/8 = +6\,kNm.$

$DF_{BA} = (3/4)(EI/6)/[(3/4)(EI/6)+(3/2)(EI/6)] = 1/3.$

For bending moment and shear force diagrams see Fig. 5.22.

5 See Fig. 5.23.

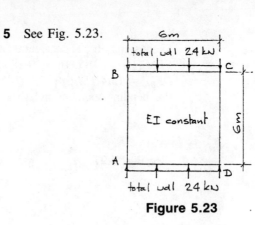

Figure 5.23

Solution:

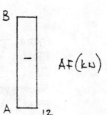

Figure 5.24

Solution

Symmetry about vertical and horizontal axes.

$\text{FEM}_{BC} = -WL/12 = -12 \text{ kNm}$.

$DF_{BA} = 1/2$.

For bending moment, shear force, and axial force diagrams see Fig. 5.24.

6 See Fig. 5.25.

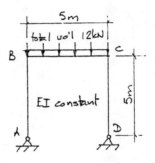

Figure 5.25

Solution:

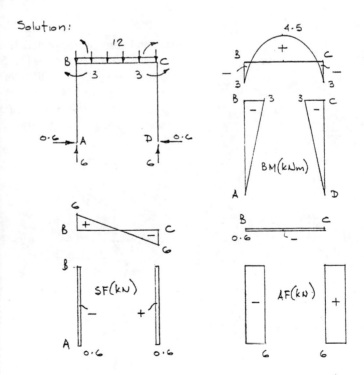

Figure 5.26

Solution

Symmetry about centre of BC.
$\text{FEM}_{BC} = -WL/12 = -5$ kNm.
$DF_{BA} = (3/4)(EI/5)/[(3/4)(EI/5)+(1/2)(EI/5)] = 3/5$.
For bending moment, shear force, and axial force diagrams see Fig. 5.26.

7 See Fig. 5.27.

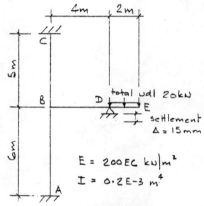

Figure 5.27

Solution

Horizontal force at support D holds B in position
$\text{FEM}_{BD} = -3EI\Delta/L^2 = -112.5$ kNm;
$\text{FEM}_{DE} = -WL/2 = -20$ kNm.
$DF_{BD} = (3/4)(EI/4)/[(3/4)(EI/4)+EI/5+EI/6] = 0.338;$
$DF_{BC} = (EI/5)/[(3/4)(EI/4)+EI/5+EI/6] = 0.361.$
For bending moment, shear force, and axial force diagrams see Fig. 5.28.

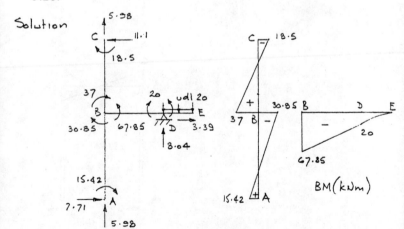

Figure 5.28

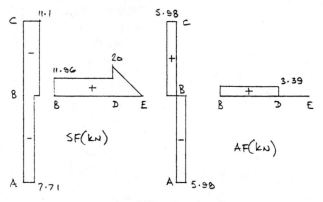

Fig. 5.28

8 See Fig. 5.29.

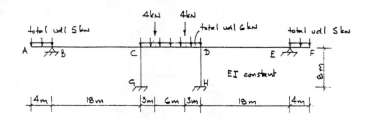

Figure 5.29

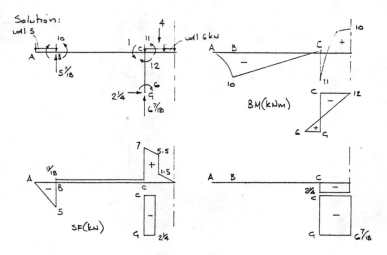

Figure 5.30

Solution

Symmetry about centre of CD.

$\text{FEM}_{BA} = +WL/2 = +10 \text{ kNm};$

$\text{FEM}_{CD} = -wL^2/12 - 3(2W)L/32 = -15 \text{ kNm}.$

$DF_{CB} = (3/4)(EI/18)/[(3/4)(EI/18)+(1/2)(EI/12)+EI/8] = 1/5$
$DF_{CD} = (1/2)(EI/12)/[(3/4)(EI/18)+(1/2)(EI/12)+EI/8] = 1/5.$
For bending moment, shear force, and axial force diagrams see Fig. 5.30.

9 See Fig. 5.31.

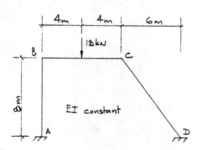

Figure 5.31

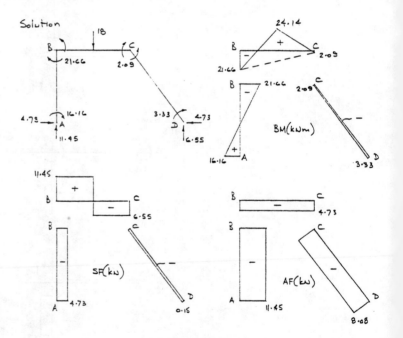

Figure 5.32

Solution
$FEM_{BC} = -WL/8 = -18\,kNm$, $FEM_{CB} = +18\,kNm$.
$DF_{BA} = EI/8/[EI/8+EI/8] = 1/2$;
$DF_{CB} = EI/8/[EI/8+EI/10] = 5/9$.

End moments after the first stage, AB($+6.06$), BA($+12.30$), BC(-12.30), CB($+10.69$), CD(-10.69), DC(-5.27) kNm.

Prop force at B in first stage is $P = 6.26$ kN.

Ratio of fixed end moments for displacements in stage two, M_{AB} : M_{BC} : M_{DC} = $1/8^2 : (3/4)/8^2 : (5/4)/10^2 = 100:75:80$.

End moments after the second stage, AB($+10.1$), BA($+9.35$), BC(-9.35), CB(-8.6), CD($+8.6$), DC($+8.6$) kNm.

For bending moment, shear force, and axial force diagrams see Fig. 5.32.

6

Deflections by the unit load method

Introduction

The unit load method, which is applicable to both pin-jointed and rigidly jointed structures, may be derived from the energy theorems of Engesser and Castigliano, or from the principle of virtual work. Either approach results in the same general formula.

Consider the plane, pin-jointed framed in Fig. 6.1. At the joints the external loads are W_1, W_2, ..., W_i, ..., W_n and the corresponding vectorially equivalent deflections are δ_1, δ_2, ..., δ_i, ..., δ_n. The complementary work done at the joints is equal to the complementary energy in the members, so, providing there is no settlement of the supports

$$\sum_1^n \int \delta_i \, dW_i = C \tag{6.1}$$

Differentiating partially with respect to one of the external loads

$$\delta_i = \partial C / \partial W_i \tag{6.2}$$

That is: *if the total complementary energy is partially differentiated with respect to an external load the result is the displacement at the load point in the direction of the load.* This is Engesser's First Theorem of Complementary Energy. It can be applied generally to pin-jointed and rigidly jointed structures, structures in which the relationship between stress and strain is either linear or non-linear, and also to inelastic structures provided that no unloading of any of the members occurs during the application of the external loads.

In the special case of a linear elastic structure strain and complementary energies are equal and may be exchanged in equation (6.2), to give

$$\delta_i = \partial U / \partial W_i \tag{6.3}$$

This is Part II of Castigliano's First Theorem.

Returning to the frame in Fig. 6.1 and letting the force in a typical member be P, the total complementary energy in the members is given by

$$C = \sum_{\text{members}} P^2 L / 2EA \tag{6.4}$$

Differentiating and substituting into equation (6.2) gives

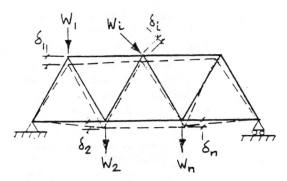

Figure 6.1

$$1 \times \delta_i = \Sigma(PL/EA)\partial P/\partial W_i \qquad (6.5)$$

Now $\partial P/\partial W_i$ is the force that would be produced in a typical member if W_i were replaced by a unit load and all the other external loads were zero. Denoting it by p and rearranging

$$\delta_i = \Sigma pPL/EA \qquad (6.6)$$

The same formula can also be derived from virtual work by applying a single unit virtual load at joint i in the direction of the required deflection δ_i, to give the typical member force p. Under the actual loading let the extension of a typical member be e. Since the actual displacements e are compatible with the actual deflections δ at the joints, and the member forces p are in equilibrium with the unit virtual load, the virtual work equation (1.5) becomes

$$0 \times \delta_1 + 0 \times \delta_2 + \ldots + 1 \times \delta_i + \ldots + 0 \times \delta_n = \underset{\text{members}}{\Sigma pe} \qquad (6.7)$$

and the substitution of PL/EA for e gives equation (6.6).

Summarising, the frame is first analysed under the actual external loads to obtain the member forces P. For each deflection it is then re-analysed under a single unit load at the point where the deflection is required. The unit load must be vectorially equivalent to, i.e. in the same sense and direction as, that assumed for the deflection. In rigidly jointed structures and cases where the loads are applied at points other than the joints, bending moments and other stress resultants will also be present. Since these are usually variable it is necessary to integrate along the length of the members. The integrands are derived by partial differentiation of the expressions in equation (1.4), thus

Axial force:	$pP\mathrm{d}z/AE$
Bending moment:	$mM\mathrm{d}z/EI$
Shear force:	$sS\mathrm{d}z/GA$
Torque:	$tT\mathrm{d}z/GJ$

$$(6.8)$$

the lower case symbol denoting the stress resultant due to the unit load. The total deflection is then given by

$$\delta_i = \Sigma \int pP\mathrm{d}z/EA + \Sigma \int mM\mathrm{d}z/EI + \Sigma \int sS\mathrm{d}z/GA + \Sigma \int tT\mathrm{d}z/GA \qquad (6.9)$$

In plane structures bending usually predominates, but torsional deformations may be significant if any of the loads are applied at an angle to the plane of the structure. In braced arches and similar types of structure it may be necessary to include the effects of axial forces. Shear force, however, has no significant effect on deformations except in members that are very deep in relation to their length; and in such cases it is likely that methods of analysis other than the simple theory of bending would need to be considered.

Thermal strains

In a statically determinate structure thermal straining produces no stress in the members. However, it does change the size, and possibly the shape, of the structure. In a pin-jointed frame, assuming λ to be the thermal expansion of a typical member, the total complementary energy from equation (1.3) is

$$C = \Sigma(P\lambda + P^2L/2EA)$$

and $\delta_i = \partial C/\partial W_i = \Sigma(\lambda + PL/EA)\partial P/\partial W_i$

i.e. $\delta_i = \Sigma p(\lambda + e)$ (6.10)

where $e = PL/EA$ and $\lambda + e$ is therefore the total extension of the member.

Worked examples

6.1 Pin-jointed frame

Find the vertical and horizontal deflections at joint D in the pin-jointed plane frame of Fig. 6.2. All members have cross sectional area 1000 mm^2 and Young's modulus 200 kN/mm^2.

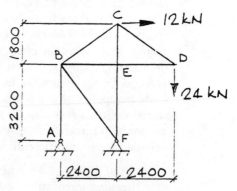

Figure 6.2

Solution The solution consists of the following stages.

(1) Find the member forces P due to the general loading, using any of the standard methods described in Volume 1 of *Solving Problems in Structures*. Sign convention: tension positive.

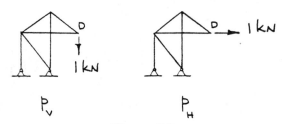

Figure 6.3

(2) Find the member forces p_V due to a unit vertical load at joint D.
(3) Find the member forces p_H due to a unit horizontal load at D.
(4) For each member calculate $p_V PL/EA$ and $p_H PL/EA$.
(5) Use equation (6.6), i.e. $\delta_V = \Sigma p_V PL/EA$ and $\delta_H = \Sigma p_H PL/EA$. If the result is negative, the deflection is in the opposite direction to the unit load.

After determining the member forces computation is best carried out in tabular form as follows. The units are kN and mm.

Member	P	L	PL/EA	$\cdot\ p_V$	p_H	$p_V PL/EA$	$p_H PL/EA$
AB	49	3200	0.784	1	4/3	0.784	1.045
BC	55	3000	0.825	5/3	0	1.375	0
CD	40	3000	0.600	5/3	0	1.000	0
DE	−32	2400	−0.384	−4/3	1	0.512	−0.384
BE	−32	2400	−0.384	−4/3	1	0.512	−0.384
CE	−57	1800	−0.513	−2	0	1.026	0
EF	−57	3200	−0.912	−2	0	1.824	0
BF	−20	4000	−0.400	0	−5/3	0	0.667
					Σ	7.033	0.944

Hence the deflections are 7.033 mm vertically downwards and 0.944 mm horizontally to the right.

6.2 Simply supported beam with distributed load

Show that the central deflection of a simply supported beam with a uniformly distributed load is $5wL^4/384EI$.

Solution − Fig. 6.4 Using the bending moment expression of equation (6.8) and taking account of symmetry,

$$\delta = 2 \int_0^{L/2} mM\,dz/EI$$

where, for $z \leq L/2$,

$$M = wLz/2 - wz^2/2, \quad m = z/2$$

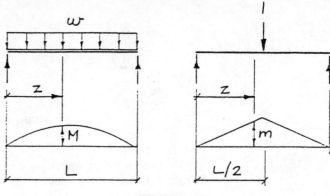

Figure 6.4

Hence

$$\delta = w\int(Lz^2 - z^3)dz/2EI = 5wL^4/384EI$$

6.3 Cantilever with point loads — use of superposition

Find the deflection at the free end of the cantilever in Fig. 6.5(a). Hence find the deflection for the multiple load-case in Fig. 6.5(b) when $EI = 6250\ kNm^2$.

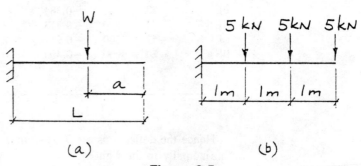

(a)   (b)

Figure 6.5

Solution — Fig. 6.6

	M	m	mM
$z = 0$ to a	0	$-z$	0
$z = a$ to L	$-W(z-a)$	$-z$	$Wz(z-a)$

$$\delta = \Sigma\int mM\,dz = W\int_a^L (z^2 - az)dz/EI = W(2L^3 - 3aL^2 + a^3)/6EI$$

Note that z can be measured from any convenient point, provided that it is the same for both m and M.

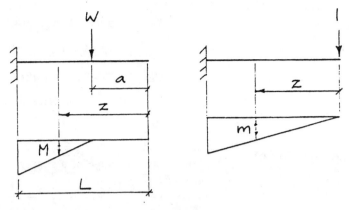

Figure 6.6

To find the deflection in Fig. 6.5(b) put $W = 5\,\text{kN}$, $L = 3\,\text{m}$, and add the results for $a = 0, 1, 2\,\text{m}$.

Hence

$$\delta = 5(54+28+8)/(6\times6250) = 0.012\,\text{m}$$

6.4 Simply supported beam with point loads

Derive an expression for the cental deflection of a simply supported beam with a point load a distance a from the left hand support ($a \leq L/2$).

Solution — Fig. 6.7 As the discontinuities in the bending moment diagrams for M and m do not coincide, integration must be taken over three parts of the beam, as follows.

	M	m	mM
$z = 0$ to a	$Wz(L-a)/L$	$z/2$	$Wz^2(L-a)/2L$
$z = a$ to $L/2$	$Wa(L-z)/L$	$z/2$	$Waz(L-z)/2L$
$z = L/2$ to L	$Wa(L-z)/L$	$(L-z)/2$	$Wa(L-z)^2/2L$

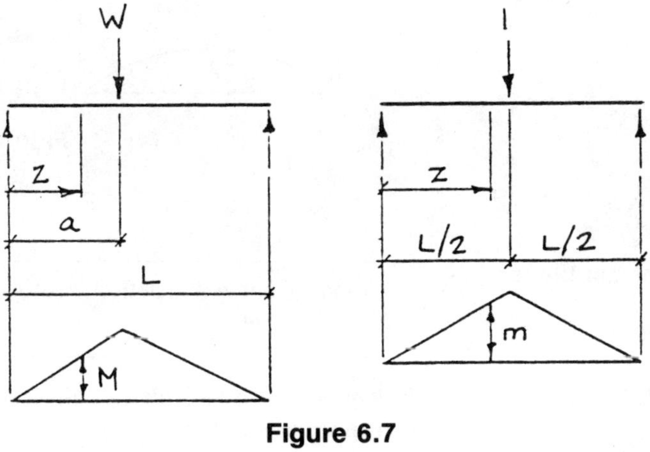

Figure 6.7

$$\delta = W\left\{\int_0^a (Lz^2 - az^2)\mathrm{d}z + \int_a^{L/2}(aLz - az^2)\mathrm{d}z\right.$$
$$\left. + \int_{L/2}^{L}(aL^2 - 2aLz + az^2)\mathrm{d}z\right\}/2LEI$$
$$= WL^3\{3a/L - 4(a/L)^3\}/48EI$$

Compare this solution with Macaulay's method in Chapter 2, Problem 1.

6.5 Determination of slope

Show that the slope of a simply supported beam at the supports is $wL^3/24EI$.

Solution — Fig. 6.8 The slope of a beam is a measure of its rotation, so that in order to be vectorially equivalent the unit load must be a couple. Otherwise the procedure is the same as for other beam problems. This problem can be simplified slightly by taking symmetry into account and applying two unit couples as shown, then integrating over half the span.

$$M = wLz/2 - wz^2/2, \quad m = 1$$

and

$$\theta = \Sigma \int mM\mathrm{d}z/EI = w\int_0^{L/2}(Lz - z^2)\mathrm{d}z/2EI = wL^3/24EI$$

Check the result by applying a unit couple at one end only and integrating over the whole span.

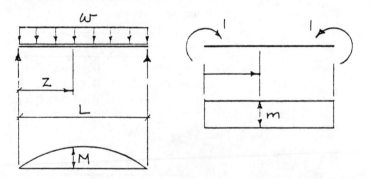

Figure 6.8

6.6 Bent cantilever

Find the horizontal deflection at A in the bent cantilever of Fig. 6.9(a). *EI* is constant.

Solution — Fig. 6.9(b) If the unit is applied to the right as shown, the bending moments are as follows.

	M	m	mM
AB	0	z	0
BC	$-wz^2/2$	h	$-whz^2/2$

$$\delta = \Sigma \int mM\,dz/EI = -wh\int_0^L z^2 dz/2EI = -whL^3/6EI$$

The negative sign indicates that the deflection is to the left, opposing the unit load.

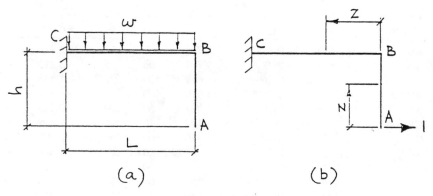

Figure 6.9

6.7 Effect of thermal strain

In the frame of Fig. 6.10 the members are all 3 m long. The top members are ornamental as well as structural and are made from an alloy, while the others are light steel sections. In addition to the loading shown the frame is subjected to a uniform rise in temperature of 15°C. Cross sectional areas, Young's moduli, and coefficients of linear expansion are as follows.

	$A\,(\text{mm}^2)$	$E\,(\text{kN/mm}^2)$	$\alpha/°C$
AB,BD,DF,FG	100	75	$19E-6$
Others	60	200	$12E-6$

Determine:
(a) the vertical deflection at joints C and E
(b) the vertical deflections at D,C, and E due to the rise in temperature alone if all the members were steel.

Solution (a) The method of solution is as follows.

(1) Determine the member forces due to the external loads and to unit vertical loads at C and E. Let these be respectively P, p_C, and p_E. In this case the work can be shortened if the forces p_C are found first. Then, by skew symmetry, $p_E(\text{EF}) = p_C(\text{BC})$, $p_E(\text{AB}) = p_C(\text{FG})$, and so on. Finally $P = 6p_C + 3p_E$.

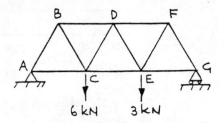

Figure 6.10

Member	p_C	p_E	P	A	E	$e=PL/EA$	λ	$p_C(\lambda+e)$	$p_E(\lambda+e)$
AB	−0.7698	−0.3849	−5.774	100	75	−2.310	0.855	1.120	0.560
BD	−0.7698	−0.3849	−5.774	100	75	−2.310	0.855	1.120	0.560
DF	−0.3849	−0.7698	−4.619	100	75	−1.848	0.855	0.382	0.764
FG	−0.3849	−0.7698	−4.619	100	75	−1.848	0.855	0.382	0.764
AC	0.3849	0.1925	2.887	60	200	0.719	0.540	0.485	0.242
CE	0.5774	0.5774	5.197	60	200	1.299	0.540	1.062	1.062
EG	0.1925	0.3849	2.310	60	200	0.576	0.540	0.215	0.430
BC	0.7698	0.3849	5.774	60	200	1.444	0.540	1.527	0.764
CD	0.3849	−0.3849	1.155	60	200	0.289	0.540	0.319	−0.319
DE	−0.3849	0.3849	−1.155	60	200	−0.289	0.540	−0.097	0.097
EF	0.3849	0.7698	4.619	60	200	1.155	0.540	0.652	1.305
							Σ	7.167	6.229

(2) Use a tabular computation to evaluate equation (6.10). The units are kN and mm.

i.e. $\delta_C = 7.167\,\text{mm}$, $\delta_E = 6.229\,\text{mm}$

Solution (b) The same method of solution could be used here, simply by applying an additional unit load at D and omitting the columns for *P, A, E,* and *e*, but as all the members are now of the same material the expansion is uniform. Hence, as the structure is simply supported, it simply increases in size without changing its shape.

i.e. $\delta_D = \lambda_{\text{steel}}\sin 60° = 0.540\sin 60° = 0.468\,\text{mm}$ (upwards)
and $\delta_C = \delta_E = 0$

Tabulation of integrals

It can be shown that the value of the integral of the product of two bending moment expressions is only dependent on the shapes of the bending moment diagrams and their maximum ordinates. If *a* and *c* are the maximum ordinates of two expressions m_i and m_j respectively,

$$\int_0^L m_i m_j \, dz = \alpha Lac$$

where α is a coefficient equal to the value of the integral when both ordinates are unity.

Table 6.1

$m_i \backslash m_j$	▭ (c)	◣	◢	△	◠	◰	◿	◪	◹
▭ (a)	1	1/2	1/2	1/2	2/3	1/3	1/3	2/3	2/3
◣	1/2	1/3	1/6	1/4	1/3	1/4	1/12	5/12	1/4
◢	1/2	1/6	1/3	1/4	1/3	1/12	1/4	1/4	5/12
△	1/2	1/4	1/4	1/3	5/12	7/48	7/48	17/48	17/48
◠	2/3	1/3	1/3	5/12	8/15	1/5	1/5	7/15	7/15
◰	1/3	1/4	1/12	7/48	1/5	1/5	1/30	3/10	2/15
◿	1/3	1/12	1/4	7/48	1/5	1/30	1/5	2/15	3/10
◪	2/3	5/12	1/4	17/48	7/15	3/10	2/15	8/15	11/30
◹	2/3	1/4	5/12	17/48	7/15	2/15	3/10	11/30	8/15

Values of α for combinations of diagrams resulting from simple applications of uniformly distributed and point loads are given in Table 6.1. The idea of tabulation was proposed by Morice (1959). With a little practice rapid solutions can be obtained in cases such as Example 6.8 where the bending moment diagrams for m and M have the same base length and are continuous functions. When diagrams overlap, i.e. when their base lengths are different or, as in Example 6.9, where discontinuities are not coincident, they must be broken up into smaller units and the advantage over direct integration is reduced.

Worked examples

6.8 Use of tabulated integrals

Determine the horizontal and vertical deflections at the free end C of the cantilever in Fig. 6.11, using the table of integrals. Young's modulus E is constant and the second moments of area are as shown.

Solution — Fig. 6.12 (1) Draw the bending moment diagram (M) for the actual loading.

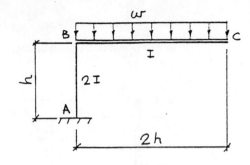

Figure 6.11

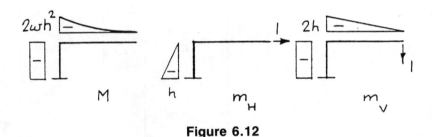

Figure 6.12

(2) Apply unit horizontal and vertical loads at C and draw the bending moment diagrams (m_H and m_V).

(3) The horizontal deflection is given by

$$\delta_H = [\int m_H M \, dz/2EI]_{AB} + [\int m_H M \, dz/EI]_{BC} \tag{1}$$

which, as the diagrams do not overlap, may be written in terms of the symbols of Table 6.1 thus

$$\delta_H = [\alpha Lac/2EI]_{AB} + [\alpha Lac/EI]_{BC} \tag{2}$$

Enter Table 6.1 with the shapes of the bending moment diagrams.
For AB: $m_H M = \triangleright \times \square$ for which $\alpha = 1/2$ (row 2, column 1) and the length $L = h$. From Fig. 6.12 the maximum values of the bending moments are $a = -h$, $c = -2wh^2$. On BC there is no bending moment m_H, so this member makes no contribution to the horizontal deflection. Substitution into equation (2) therefore gives

$$\delta_H = wh^4/2EI$$

(4) Vertical deflection is obtained simply by substituting m_V for m_H in equation (1), as follows

	α	L	a	c
AB: $m_V M = \square \times \square$	1	1	$-2h$	$-2wh^2$
BC: $m_V M = \triangleright \times \triangleleft$	1/4	2h	$-2h$	$-2wh^2$

Hence, from equation (2),

$$\delta_V = 4wh^4/EI$$

6.9 Overlapping bending moment diagrams

A simply supported beam with a span of 16 m carries a point load W at the quarter-span position. If EI is constant, determine the central deflection.

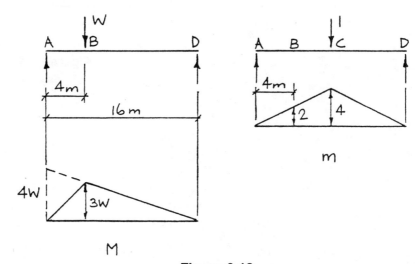

Figure 6.13

Solution — Fig. 6.13 The maximum bending moments are $M = 3W$ at B and $m = 4$ at C. As these values occur at different points it is necessary to perform the integration in stages. There are a number of ways in which this can be done. The following, which requires only two stages, considers the diagram for M as the difference between two triangles with bases AD and AB. The first is multiplied by m from A to D, the second by m from A to B.

		α	L	a	c
AD: $mM =$	△ × ◿	1/4	16	4	$4W$
AB: $mM =$ − ◿ × ◿		1/6	2	2	$4W$

$$\Sigma \alpha L a c = 1/4 \times 16 \times 4 \times 4W - 1/6 \times 2 \times 2 \times 2W = 58.7W$$

Hence

$$\delta_C = 58.7W/EI$$

Compare this solution with that of Example 6.4. When the dimensions are given in general terms solution by direct integration is often easier than by tabulated integrals.

Problems

1 Derive an expression for the deflection at the end of the stepped cantilever in Fig. 6.14 using the unit load method.

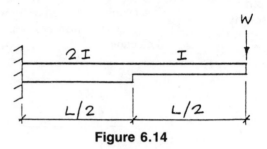

Figure 6.14

Solution

$M = -Wz$, $m = -z$

$\delta = \int mM\,dz/EI = W\Big|_0^{L/2} z^2 dz/EI + W\Big|_{L/2}^{L} z^2 dz/2EI = 3WL^3/16EI$

2 Find the vertical deflection at C of the pin-jointed frame in Fig. 6.15. The members are of steel with Young's modulus 200 kN/mm² and have the following cross sectional areas: AD and DE 150 mm², all others 50 mm².

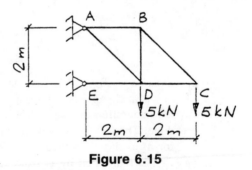

Figure 6.15

Solution

Member forces (kN) *P, p*
AB 5,1; BC 5√2, √2; CD −5, −1; DE −15, −2; AD 10√2, √2;
BD −5, −1
$\delta = \Sigma pPL/EA = 9.72$ mm

3 Derive expressions for the vertical and horizontal deflections at the free end of the bent cantilevers in Fig. 6.16.

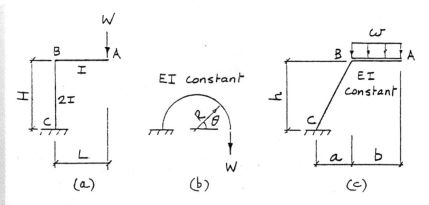

Figure 6.16

Solution

(a) Bending moments M, m_V, m_H (unit load to right)

AB $-Wz$, $-Z$, 0; BC $-WL$, $-L$, $-z$

$$\delta_V = W\int_0^L z^2 dz/EI + W\int_0^H L^2 dz/2EI = WL^2(2L+3H)/6EI$$

$$\delta_H = W\int_0^H Lz dz/EI = WLH^2/2EI \text{ (to right)}$$

(b) $M = -WR(1-\cos\theta)$, $dz = R\,d\theta$

$m_V = -R(1-\cos\theta)$, m_H (unit load to right) $= R\sin\theta$

$$\delta_V = WR^3\int_0^\pi (1-\cos\theta)^2 d\theta/EI = 3WR^3\pi/2EI$$

$$\delta_H = -WR^3\int_0^\pi (1-\cos\theta)\sin\theta\, d\theta/EI = -2WR^3/EI \text{ (to left)}$$

(c) z measured: in AB from A to B, in BC from B to C

Bending moments M, m_v, m_H (to right)

AB $-wz^2/2$, $-z$, 0

BC $-(wb^2/2 + wbz\cos\theta)$, $-(b + z\cos\theta)$, $-z\sin\theta$

$\cos\theta = a/\sqrt{(a^2+h^2)}$

$$\delta_V = w\int_0^b z^3 dz/2EI + wb\int_0^{a\sec\theta}(b^2 + 3zb\cos\theta + 2z^2\cos^2\theta)dz/2EI$$

$$= w\{3b^4 + 2b[\sqrt{(a^2+h^2)}][4a^2+9ab+6b^2]\}/24EI$$

$$\delta_H = wb\sin\theta\int_0^{a\sec\theta}(bz + 2z^2\cos\theta)dz/2EI$$

$$= wbh[\sqrt{(a^2+h^2)}][4a+3b]/12EI \text{ (to right)}$$

4 Determine the vertical deflection at the free end A of the cantilever ABC in Fig. 6.17. The whole beam lies in a horizontal plane and carries a vertical load of 3 kN at A. The section is solid circular, 80 mm in diameter. Young's modulus is 200E3 N/mm² and the shear modulus is 80E3 N/mm².

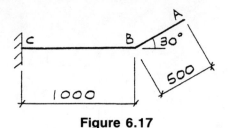

Figure 6.17

Solution

$EI = 401.1E6 \text{ kNmm}^2$, $GJ = 321.7E6 \text{ kNmm}^2$

Bending moments M, m (measuring z from A in AB and B in BC)

AB $-3z$, $-z$; BC $-3(500\cos30° + z)$, $-(500\cos30° + z)$

Torques T, t

AB 0, 0; BC $3 \times 500\sin30°$, $500\sin30°$

$$\delta = 3\int_0^{500} z^2 dz/EI + 3\int_0^{1000} (500\cos30° + z)^2 dz/EI$$

$$+ 3\int_0^{1000} (500\sin30°)^2 dz/GJ = 8.03 \text{ mm}$$

5 Derive expressions for the deflection at the free ends and the slope at the supports of the beam in Fig. 6.18. Assume that the section is uniform.

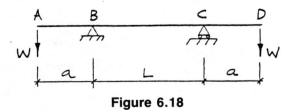

Figure 6.18

Solution

Make use of symmetry by applying a unit vertical force at each free end for m_V and a unit hogging moment at each support for m_θ; then consider half of the beam.

	M	m_V	m_θ
AB	$-Wz$	$-z$	0
BC (to mid-span)	$-Wa$	$-a$	-1

$$\delta_A = W\int_0^a z^2 dz/EI + W\int_0^{L/2} a^2 dz/EI = Wa^2(2a + 3L)/6EI$$

$$\theta_B = W\int_0^{L/2} a\, dz/EI = WaL/2EI$$

6 Use the table of integrals to determine the central deflection normal to the beam in Fig. 6.19. EI is constant.

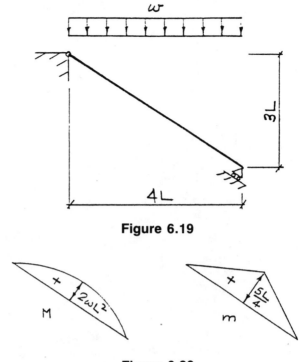

Figure 6.19

Figure 6.20

Solution − Fig. 6.20

For *m* the unit load is applied normal to the beam

$$\delta = \Sigma \alpha Lac/EI = 5/12 \times 5L \times 2wL^2 = 125wL^4/24EI$$

7   The beam in Fig. 6.21 has a uniform section and is continuous over simple supports at B and D. Use the table of integrals to determine the value of the load P if the deflections at A and E are to be zero. What is the deflection at C in the centre of the beam with this loading?

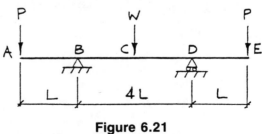

Figure 6.21

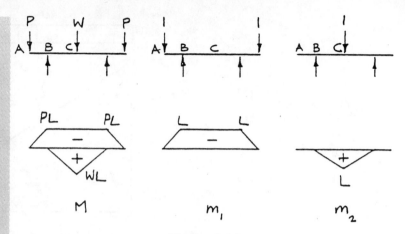

Figure 6.22

Solution − Fig. 6.22

   (1) Draw the bending moment diagrams for the external loads P and W separately.

   (2) For the end deflections apply two unit loads to maintain symmetry; then consider half the beam.

		α	L	a	c
$(m_1M)_{AB} =$	△×△	1/3	L	$-L$	$-PL$
$(m_1M)_{BC} =$	□×□	1	$2L$	$-L$	$-PL$
$+$	□×△	1/2	$2L$	$-L$	WL

$EI\delta_A = \Sigma\alpha Lac = PL^3/3 + 2PL^3 - WL^3 = 0$, hence $P = 3W/7$

   (3) Apply a unit load at C for the central deflection.

		α	L	a	c
$(m_2M)_{AB} = 0$					
$(m_2M)_{BC} =$	△×□	1/2	$4L$	L	$-PL$
$+$	△×△	1/3	$4L$	L	WL

$$\delta_C = \Sigma\alpha Lac/EI = (-2PL^3 + 4WL^3/3)/EI$$
$$= 10WL^3/21EI \text{ (substituting for } P)$$

8 (a) Find the vertical deflection at joint B in the pin-pointed frame of Fig. 6.23. Members AB and BC are timber with $E = 5\,kN/mm^2$, the others steel with $E = 200\,kN/mm^2$. The sections are solid square with sides 300 mm and 50 mm respectively.

   (b) What additional deflection would be produced by drying out of the timber members assuming that shrinkage is 0.1% of their length?

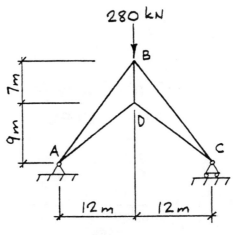

Figure 6.23

Solution

Member	p_A (down)	p_C (right)	P (kN)	L(mm)	EA (kN)
AB, BC	$-10/7$	$-15/7$	-400	20E3	450E3
AD, DC	$15/4$	$20/7$	300	15E3	500E3
BD	$9/7$	$24/7$	360	7E3	500E3

(a) $\delta_B = \Sigma p_B PL/EA = 76.6\,\text{mm}, \quad \delta_C = \Sigma p_C PL/EA = 144.9\,\text{mm}$

(b) Shrinkage $= 20\text{E}3 \times 0.001 = 20\,\text{mm}$, i.e. $\lambda = -20\,\text{mm}$

$\delta_B = \Sigma p_B \lambda = 2 \times (-10/7) \times (-20) = 57.1\,\text{mm (downwards)}$

$\delta_C = \Sigma p_C \lambda = 2 \times (-15/7) \times (-20) = 85.7\,\text{mm (to right)}$

Reference

Morice P.B. *Linear Structural Analysis*, Thames and Hudson, London, 1959.

7

Unit load method for hyperstatic structures

Introduction

The unit load method for hyperstatic structures is an extension of the method for deflections described in the previous chapter. It is an indirect flexibility method which may be derived either from energy or virtual work. The relevant energy theorems are the compatibility theorems of Engesser or Castigliano which, for a structure containing n redundancies, may be stated as

$$\partial C/\partial R_i = 0 \tag{7.1}$$

and $\quad \partial U/\partial R_i + \lambda_i = 0 \tag{7.2}$

where $i = 1,2,\ldots,n$ and λ_i is the lack of fit in a redundant member.

In this chapter Engesser's theorem has been used because of its greater generality — Castigliano's theorem is restricted to linear elastic structures with no initial deformations. A more general derivation from virtual work is given in Chapter 10.

Pin-jointed frames

Consider the pin-jointed frame in Fig. 7.1(a). There are two redundancies which may be conveniently identified as the force R_1 in member BC and the vertical reaction R_2 at joint D. Engesser's theorem therefore produces two compatibility equations $\partial C/\partial R_1 = 0$ and $\partial C/\partial R_2 = 0$. Substituting for C in equation (6.4), differentiating partially and letting $r_1 = \partial C/\partial R_1$ and $r_2 = \partial C/\partial R_2$

$$\Sigma r_1 PL/EA = 0 \tag{7.3}$$

$$\Sigma r_2 PL/EA = 0 \tag{7.4}$$

In these equations P is the force in a typical member of the hyperstatic structure and can be expressed in the form

$$P = P_w + r_1 R_1 + r_2 R_2 \tag{7.5}$$

where P_w is the member force produced by the external loads when the redundancies R_1 and R_2 are set to zero. r_1 and r_2 are influence coefficients,

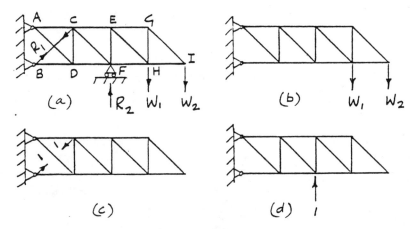

Figure 7.1

i.e. r_1 is the force produced in a typical member when a single unit force acts in place of R_1 while R_2 is set to zero and vice versa, P_w, r_1, and r_2 are, in effect, the member forces in the residual statically determinate structure obtained by removing the redundancies. They can be obtained by analysing the structures in Figs 7.1(b),(c) and (d) respectively by any of the standard methods (see Vol. 1 *Solving Problems in Structures*).

Substituting into equations (7.3) and (7.4) and transposing

$$R_1\Sigma r_1^2L/EA + R_2\Sigma r_1r_2L/EA = -\Sigma r_1P_wL/EA \qquad (7.6)$$

$$R_1\Sigma r_2r_1L/EA + R_2\Sigma r_2^2L/EA = -\Sigma r_2P_wL/EA \qquad (7.7)$$

All the members in the hyperstatic structure must be included in the summations.

The solution of equations (7.6) and (7.7) for R_1 and R_2 and back-substitution into equation (7.5) to obtain the member forces completes the analysis. The number of equations is equal to the number of redundancies and the pattern of subscripts can be inferred from equations (7.6) and (7.7). Thus in the ith equation the coefficient of the qth redundancy R_q is $\Sigma r_ir_qL/EA$ and the constant term is $-\Sigma r_iP_wL/EA$.

Lack of fit and thermal strains

When a redundant member is fitted into a stiff frame *self-straining forces* are introduced in all the members unless it is of the correct length. A similar state of affairs results after assembly when non-uniform changes in length occur due to thermal expansion etc. Problems can be dealt with, as in the previous chapter, by modifying the complementary energy to include an initial displacement λ. For pin-pointed frames the ith equation becomes

$$\Sigma r_i(\lambda + PL/EA) = 0 \qquad (7.8)$$

λ is positive when the member expands or, in cases of lack of fit, when the member is too long before assembly.

Movement at supports

When the support conditions are statically determinate, the settlement of a support merely produces a rigid-body movement of the whole structure which can be superimposed on any elastic deformation. No additional member forces are generated.

For example, suppose the support C of the beam in Fig. 7.2(a) settles by an amount δ_s and the elastic deflection at B due to external loading is δ_e, then the total deflection at B is given by

$$\delta_B = \delta_e + \delta_s a/L \tag{7.9}$$

When the support conditions are hyperstatic, as in Fig. 7.2(b), the elastic deformation of the beam is equal to the settlement δ_s. The reaction R can be determined if δ_s is known by treating it as an external load in equation (6.2).

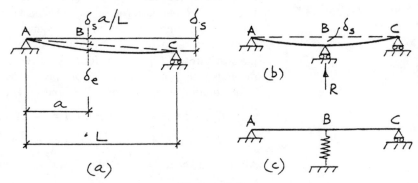

Figure 7.2

In some cases, as for example when a structure is supported by another structure, an elastic deformation may occur at a support. Neither the reaction nor the deformation are then known, but the supporting structure can usually be replaced by an elastic spring of equivalent flexibility, as in Fig. 7.2(c), and treated as an additional member of the original structure.

Flexural structures

In structures such as beams and portal frames which rely on their flexural stiffness to provide stability, the energy due to bending is usually much greater than that due to axial deformations which may therefore be ignored.

The redundant restraints can be either the reactions or the bending moments at the supports. For example, the continuous beam in Fig. 7.3(a) can be represented by either of the systems in diagrams (b) and (c). At first sight

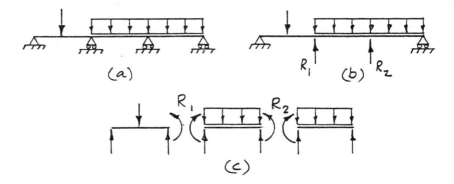

Figure 7.3

(c) may appear more complicated than (b), but in fact it produces equations that are easier to solve, especially when a large number of spans is involved. The reason is that in case (c), since each span is a free body, the influence coefficients for each redundancy apply only to the two adjoining spans; so that when compatibility equations are generated the elements of their coefficient matrix form a compact band about the leading diagonal. In case (b), on the other hand, each redundancy produces influence coefficients for every span, resulting in a full matrix. Where settlement at a support is involved, however, it is preferable to consider the reaction as redundant.

The method of analysis is basically the same as for pin-pointed frames except that integration along the length of the members is required. For a structure with n redundancies R_1, R_2, ..., R_q, ..., R_n there are n equations of the form

$$\partial C/\partial R_q = \Sigma \int M(\partial M/\partial R_q)\mathrm{d}z/EI = 0 \qquad (7.10)$$

Letting M_w be the bending moment in the statically determinate system due to the external loads, and letting $m_q = \partial M/\partial R_q$ then

$$M = M_w + m_1R_1 + \ldots + m_qR_q + \ldots + m_nR_n$$

and $\partial C/\delta R_q = \Sigma \int (M_w + m_1R_1 + \ldots + m_qR_q + \ldots + m_nR_n)m_q\,\mathrm{d}z/EI = 0$

The qth equation will therefore have the form

$$R_1\Sigma \int m_q m_1 \mathrm{d}z/EI + \ldots + R_q\Sigma \int m_q^2\,\mathrm{d}z/EI + \ldots + R_n\Sigma \int m_q m_n \mathrm{d}z/EI$$
$$= -\Sigma \int m_q M_w \mathrm{d}z/EI \qquad (7.11)$$

As in the previous chapter, the integration can be performed either directly or by means of Table 6.1. In numerical problems it is usually most convenient to choose the origin of z so that the lower limit of integration is zero.

Deflection of hyperstatic structures

Deflections of hyperstatic structures can be determined in the same general way as for statically determinate structures except that the member forces must

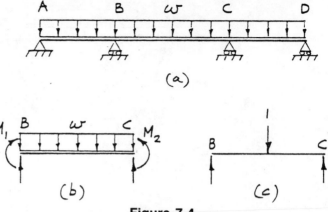

(a)

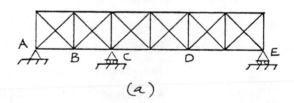

(b)   (c)

Figure 7.4

include the effects of all the redundant restraints as well as the external loads. Consider for example the continuous beam in Fig. 7.4(a), for which the central deflection of the middle span BC is required. This structure has two redundancies. Assume that they have been found; so that the moments M_1 and M_2 over the supports are known. The behaviour of span BC can now be determined by considering it as the statically determinate free body in Fig. 7.4(b) for which the deflection is given by

$$\delta = \Sigma \int mM \, dz/EI \tag{7.12}$$

where M and m are the bending moments obtained from Fig. 7.4(b) and (c) respectively.

Similarly, for a pin-jointed frame the deflection is given by

$$\delta = \Sigma pPL/EA \tag{7.13}$$

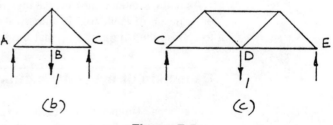

(a)

(b)   (c)

Figure 7.5

It is only necessary to consider that part of the structure containing the point of interest, and to include just enough of the existing bracing to make a statically determinate system that will effectively locate the joints between points of zero deflection. In the hyperstatic frame of Fig. 7.5(a) for example, after analysing the whole frame to determine the member forces, it is only necessary to consider the sub-structures in diagrams (b) and (c) to determine the vertical deflections of joints B and D.

Worked examples

7.1 Symmetrical pin-jointed frame with redundant reaction

Determine the reactions at A and C and the member forces in the frame of Fig. 7.6. Assume that EA is the same for all members.

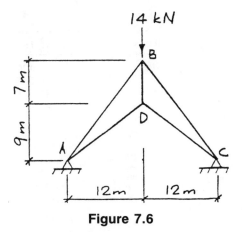

Figure 7.6

Solution — Fig. 7.7. (1) Let the horizontal reaction at A be the redundant restraint R_1.

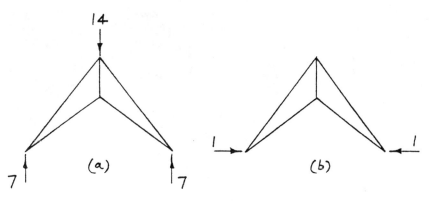

Figure 7.7

(2) Determine the member forces P_w by setting R_1 to zero, as in Fig. 7.7(a). Determine the influence coefficients r_1 by replacing R_1 with a unit load, as in Fig. 7.7(b). Because of symmetry it is only necessary to find the forces in members AB, AD, and BD.

(3) Use equation (7.3), i.e.

$$\sum r_1 PL/EA = 0$$

where $P = P_w + r_1 R_1$
Since EA is constant

$$\sum r_1 P_w L + R_1 \sum r_1^2 L = 0$$

Hence

$$R_1 = -\sum r_1 P_w L / \sum r_1^2 L$$

(4) Set up a table, using kN and m units, as follows — completing the last column after R_1 has been calculated.

Member	P_w	r_1	L	$r_1 P_w$	$r_1^2 L$	$P = P_w + r_1 R_1$
AB	−20	15/7	20	−857.1	91.8	−5.60
AD	15	−20/7	15	−642.9	122.4	−4.20
BD	18	−24/7	7	−432.0	82.3	−5.04

$$\sum r_1 P_w L = -[2(857.1 + 642.9) + 432.0] = -3432$$
$$\sum r_1^2 L = 2(91.8 + 122.4) + 82.3 = 510.7$$
$$R_1 = 3432/510.7 = 6.720 \text{ kN}$$

7.2 Symmetrical frame with two redundant members

Determine the member forces in the frame of Fig. 7.8. The panels are square and EA is constant.

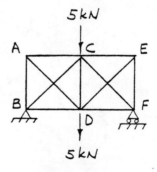

5 kN

A   C   E

B   F

D

5 kN

Figure 7.8

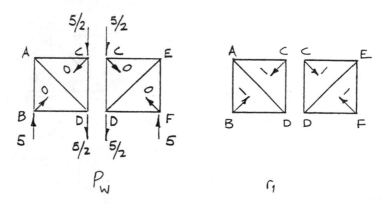

Figure 7.9

Solution − Fig. 7.9 (1) The frame has two redundant members, but as the frame and the loading are symmetrical the number of restraints to be determined is reduced to one if symmetrical members are chosen, say BC and CF. The analysis can then be carried out on half the frame. Note, however, that halving the frame necessitates cutting member CD along its longitudinal axis; so that its cross sectional area, and the external loads at C and D must be halved.

(2) Set up a table for the left hand half of the frame. Since it is only the ratio of values of L/EA that is important, let the panel length be unity. Similarly let EA be unity except for member CD where $EA = 1/2$. Determine P_w and r_1 by setting R_1 to 0 and 1 respectively, as shown.

Member	P_w	r_1	L	EA	$P=P_w+r_1R_1$
AB	−5	$-1/\sqrt{2}$	1	1	−2.26
CD	−5/2	$-1/\sqrt{2}$	1	1/2	−0.24
AC	−5	$-1/\sqrt{2}$	1	1	−2.26
BD	0	$-1/\sqrt{2}$	1	1	2.74
AD	$5\sqrt{2}$	1	$\sqrt{2}$	1	3.20
BC	0	1	$\sqrt{2}$	1	−3.87

As in the previous problem

$$R_1 = -\Sigma r_1 P_w L/EA \ / \ \Sigma r_1^2 L/EA = -20.61/5.328 = -3.868$$

(3) Back-substitute into the table to obtain the actual member force P.

i.e. $P = P_w - 3.868r_1$

7.3 Three redundancies − symmetrical loading

Determine the member forces in the symmetrical pin-pointed frame of Fig. 7.10, assuming square panels and constant EA.

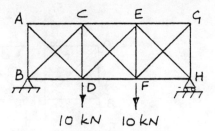

Figure 7.10

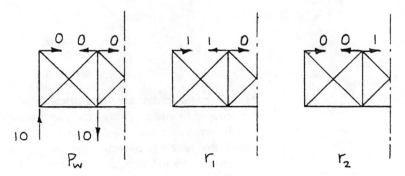

Figure 7.11

Solution — Fig. 7.11 (1) Select redundant members which, on removal, would leave a symmetrical statically determinate system: say AC, CE, EG with member forces R_1, R_2, R_3. Owing to symmetry $R_3 = R_1$.

(2) Set up the table for half the frame. In this problem all the members of the centre panel are cut. Their lengths are therefore halved, but their cross sectional areas are unaffected. Find the member forces P_w, r_1, and r_2, using the load-cases shown.

Member	P_w	r_1	r_2	L	$P = P_w + r_1 R_1 + r_2 R_2$
AB	0	1	0	1	−4.89
CD	20	1	1	1	4.05
AC	0	1	0	1	−4.89
CE	0	0	1	1/2	−11.06
BD	10	1	0	1	5.11
DF	20	0	1	1	8.94
AD	0	$-\sqrt{2}$	0	$\sqrt{2}$	6.92
BC	$-10\sqrt{2}$	$-\sqrt{2}$	0	$\sqrt{2}$	−7.23
CF	$-10\sqrt{2}$	0	$-\sqrt{2}$	$\sqrt{2}/2$	1.50
DE	$-10\sqrt{2}$	0	$-\sqrt{2}$	$\sqrt{2}/2$	1.50

(3) Use equations (7.6) and (7.7). Since EA is constant they become

$$R_1\Sigma r_1^2 L + R_2\Sigma r_1 r_2 L = -\Sigma r_1 P_w L$$
$$R_1\Sigma r_2 r_1 L + R_2\Sigma r_2^2 L = -\Sigma r_2 P_w L$$

Hence, using the values from the table

$$9.657R_1 + 1.000R_2 = -58.28$$
$$1.000R_1 + 4.828R_2 = -58.28$$

Solving, $R_1 = -4.89\,\text{kN}$, $R_2 = -11.06\,\text{kN}$

(4) Back-substitute into the table to obtain P, i.e.

$$P = P_w - 4.89r_1 - 11.06r_2$$

7.4 Three redundancies – skew-symmetrical frame

Determine the member forces in the previous problem when the load at F is reversed.

Solution – Figs 7.10 and 7.12 (1) Let members AC, CE, and EG be redundant, as in the previous problem.

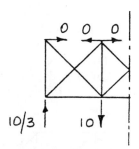

Figure 7.12

(2) Since the loading is skew-symmetrical and the frame is truly symmetrical geometrically the member forces on the right of the axis of symmetry have the same numerical values as the corresponding ones on the left, but are of opposite sign. Hence $R_3\,(F_{EG}) = -R_1\,(F_{AC})$.

All the members in the centre panel are cut by the axis of symmetry. Of these CE and DF are divided symmetrically and since the forces on either side cannot be of opposite sign they must be zero. Hence $R_2\,(F_{CE}) = 0$. The same conclusions cannot however be drawn for members CF and DE as they are not divided symmetrically.

It follows that there is only one redundancy (R_1) to be found and this can be done by considering the left hand half of the frame. The load-case for P_w is as shown in Fig. 7.12; values of r_1 are the same as for the previous problem.

(3) Set up the table, as follows.

Member	P_w	r_1	L	$P = P_w + r_1 R_1$
AB	0	1	1	−2.01
CD	20/3	1	1	4.65
AC	0	1	1	−2.01
CE	0	0	1/2	0.00
BD	10/3	1	1	1.32
DF	0	0	1/2	0.00
AD	0	$-\sqrt{2}$	$\sqrt{2}$	2.85
BC	$-10\sqrt{2}/3$	$-\sqrt{2}$	$\sqrt{2}$	−1.87
CF	$-10\sqrt{2}/3$	0	$\sqrt{2}/2$	−4.71
DE	$10\sqrt{2}/3$	0	$\sqrt{2}/2$	4.71

(4) $R_1 = -\Sigma r_1 P_w L / \Sigma r_1^2 L = -19.43/9.657 = -2.012 \text{ kN}$

(5) Back-substitute for R_1 to obtain P, i.e.

$$P = P_w - 2.012 r_1$$

7.5 Asymmetrical loading by superposition of load-cases

The frame in the previous two problems carries a single 20 kN load at D.

(a) Use the results of the previous two problems to solve the frame.

(b) Write down a set of equations from which the problem could be solved without superposition.

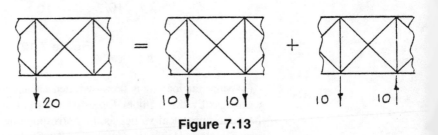

Figure 7.13

Solution − Figs 7.10 and 7.13 (a) The loading of Fig. 7.13 is obtained by superimposing the two previous loadings, as shown. Therefore it is only necessary to add the results of the two previous problems, considering both sides of the frame. The results are given in the table on page 141. Corresponding members are arranged in pairs, with the right hand member in brackets.

Solution (b) Equations (7.6) and (7.7) can be extended as follows for three redundancies.

$$R_1 \Sigma r_1^2 L/EA + R_2 \Sigma r_1 r_2 L/EA + R_3 \Sigma r_1 r_3 L/EA = -\Sigma r_1 P_w L/EA$$
$$R_1 \Sigma r_2 r_1 L/EA + R_2 \Sigma r_2^2 L/EA + R_3 \Sigma r_2 r_3 L/EA = -\Sigma r_2 P_w L/EA$$
$$R_1 \Sigma r_3 r_1 L/EA + R_2 \Sigma r_3 r_2 L/EA + R_3 \Sigma r_3^2 L/EA = -\Sigma r_3 P_w L/EA$$

Table for solution (a)

Member	P(sym)	P(skew-sym)	P(combined)
AB	−4.89	−2.01	−6.90
(GH)	−4.89	2.01	−2.88
CD	4.05	4.65	8.70
(EF)	4.05	−4.65	−0.60
AC	−4.89	−2.01	−6.90
(EG)	−4.89	2.01	−2.88
BD	5.11	1.32	6.43
(FH)	5.11	−1.32	3.79
AD	6.92	2.85	9.77
(GF)	6.92	−2.85	4.07
BC	−7.23	−1.87	−9.10
(HE)	−7.23	1.87	−5.36
CE	−11.06	0.00	−11.06
DF	8.94	0.00	8.94
CF	1.50	−4.71	−3.21
DE	1.50	4.71	6.21

7.6 Lack of fit

During assembly of the pin-jointed frame in Fig. 7.14 the member AB was fitted last and was found to be 2 mm short. Assuming EA to be 35E3 kN for all members, determine the member forces due to:
(a) self-straining,
(b) the combined effect of external loads and lack of fit.

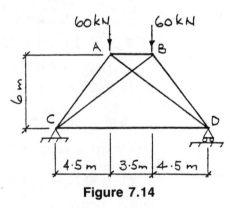

Figure 7.14

Solution − Fig. 7.15 (1) Let member AB be redundant. Determine P_w and r_1 using the load-cases shown.

(2) Construct the table, including a column for λ which in this case is the lack of fit and only applies to member AB. Since AB is too short λ is negative. Units are kN and m. Columns headed P_s and P_c are the final member forces obtained by back-substitution.

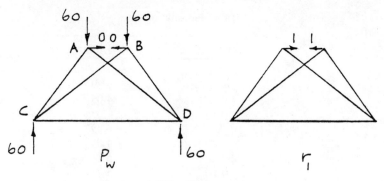

Figure 7.15

Member	P_w	r_1	λ	L	P_s	P_c
AC, BD	−48.0	0.6	0	7.5	1.85	−55.29
AD, BC	−36.0	−0.8	0	10.0	−2.47	−26.28
CD	57.6	0.28	0	12.5	0.86	54.20
AB	0.0	1.0	−0.002	3.5	3.09	−12.15

(3) Put $P = P_w + r_1 R_1$ into equation (7.8) giving

$$\Sigma r_1 \lambda + \Sigma r_1 P_w L/EA + R_1 \Sigma r_1^2 L/EA = 0$$

Hence, since EA is constant,

$$R_1 = -(\Sigma r_1 \lambda EA + \Sigma r_1 P_w L)/\Sigma r_1^2 L$$

$$\Sigma r_1 \lambda EA = -0.002 \times 35E3 = -70$$

$$\Sigma r_1 P_w L = 2(-0.6 \times 48 \times 7.5 + 0.8 \times 36 \times 10) + 0.28 \times 57.6 \times 12.5$$
$$= 345.6$$

$$\Sigma r_1^2 L = 2(0.6^2 \times 7.5 + 0.8^2 \times 10) + 0.28^2 \times 12.5 + 1^2 \times 3.5 = 22.68$$

(4) For self-straining forces put $P_w = 0$.
Hence

$$R_1 = 70/22.68 = 3.086 \text{ kN}$$

(5) For combined effects

$$R_1 = -(-70.0 + 345.6)/22.68 = -12.15$$

(6) Calculate the member forces by back-substitution in the table, i.e.

$$P_s \text{ (self-straining)} = 3.086 r_1$$
$$P_c \text{ (combined)} = P_w - 12.15 r_1$$

These calculations are completed in the table.

7.7 Settlement at a support

The frame ABCDE in Fig. 7.16 is pinned to rigid supports at A and B and rests on the mid-point of a simply supported beam at D. All the frame members have the same cross sectional area of 320 mm². The beam is 2 m long and has a second moment of area of 3.2E6 mm⁴, Young's modulus is 200 kN/mm². Determine the reaction at D:
(i) ignoring the beam deflection
(ii) taking the beam deflection into account.
By how much does support D sink in case (ii)?
How is the reaction affected if both beam supports also settle by 1 mm?

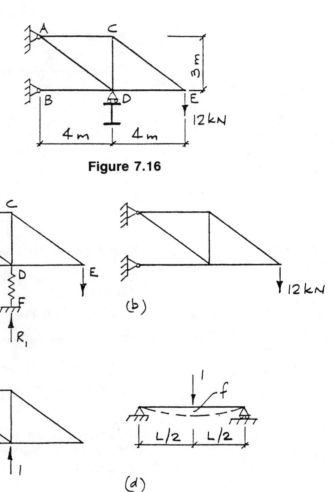

Figure 7.16

Figure 7.17

Solution — Fig. 7.17 (1) Represent the beam by an equivalent member DF, as in diagram (a).

(2) Determine P_w and r_1. For P_w R_1 is set to zero; so that the force in

member DF is also zero and the forces in the other members can be obtained from Fig. 7.17(b). Similarly the forces r_1 can be determined from Fig. 7.17(c) because R_1 and hence the force in member DF are unity.

(3) Calculate the flexibility of DF. This is the deflection of the beam produced by a unit load at the point of contact with the frame, as in diagram (d), i.e.

$$f = 1 \times L^3/48EI = 2000^3/(48 \times 200 \times 3.2E6) = 0.2604 \text{ mm/kN}$$

(4) Set up the table. The units are as shown. The axial flexibility L/EA of the frame members must be in the same units as f.

Member	P_w (kN)	r_1	L (m)	L/EA (f) (mm/kN)
AC	20	0	4	0.06255
AD	20	−5/3	5	0.07813
CE	20	0	5	0.07813
CD	−12	0	3	0.04688
BD	−32	4/3	4	0.06250
DE	−16	0	4	0.06250
DF	0	1	−	0.26040

(5) $R_1 = -\Sigma r_1 P_w L/EA \,/\, \Sigma r_1^2 L/EA$

$\Sigma r_1 P_w L/EA = -5.271$ and $\Sigma r_1^2 L/EA = 0.5885$ (including member DF)

(i) When the deflection of the beam is ignored, the terms for member DF are ignored. Hence

$R_1 = 5.271/(0.5885 - 0.2604) = 16.07 \text{ kN}$

(ii) When the beam deflection is included

$R_1 = 5.271/0.5885 = 8.956 \text{ kN}$

(6) Support D sinks by the amount that the beam deflects,

$\delta = \text{load} \times \text{flexibility} = 8.956 \times 0.2604 = 2.33 \text{ mm}$

(7) When the beam supports settle, the effect is as if point F settles by 1 mm, i.e. $\partial C/\partial R_1 = -1$ mm. The sign is negative because the direction is opposite to the direction of the unit load.
Hence

$$\partial C/\partial R_1 = \Sigma r_1 (P_w + r_1 R_1) L/EA = -1$$

so $\quad R_1 = -(1 + \Sigma r_1 P_w L/EA) \,/\, \Sigma r_1^2 L/EA$
$\quad\quad = -(1-5.271)/0.5885 = 7.257 \text{ kN}$

Note: Care is needed with signs in this problem and it is useful to remember that when settlement occurs, the corresponding redundant reaction is reduced.

7.8 Propped cantilever

Assuming that EI is constant and there is no settlement, find the reaction at B to the cantilever in Fig. 7.18.

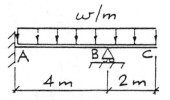

Figure 7.18

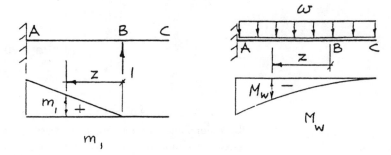

Figure 7.19

Solution $-$ *Fig. 7.19* Let R_1 be the reaction at B. As this is the only redundant restraint equation (7.11) reduces to

$$R_1\sum\int m_1^2 dz/EI = -\sum\int m_1 M_w dz/EI$$

where m_1 is the bending moment produced by a unit force at B and M_w is the bending moment due to the external loads with R_1 set to zero, i.e. with the support at B removed. From Fig. 7.19

C to B $m_1 = 0$
B to A $m_1 = z$, $M_w = -w(z+2)^2/2$

Hence
$$R_1\int_0^4 z^2 dz/EI = -w/2 \times \int_0^4 (z^3 + 4z^2 + 4z)dz/EI$$

Integrating and solving,

$$R_1 = 17w/4$$

7.9 Symmetrical pinned-base portal

A rectangular portal frame of height H and span L has pinned bases and carries a central point load W. The second moment of area of the beam is twice that of the columns and Young's modulus is constant. Find the bending moments at the eaves and at mid-span.

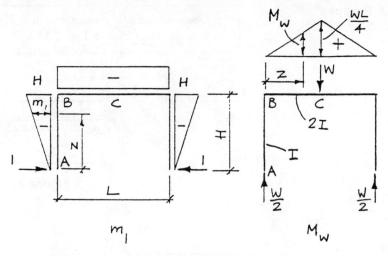

Figure 7.20

Solution — Fig. 7.20   (1) Letting the horizontal reaction at the bases be R_1, the redundant restraint, obtain expressions for m_1 and M_w from Fig. 7.20. Taking account of symmetry and considering half the frame, say ABC,

AB: $m_1 = -z$, $M_w = 0$
BC: $m_1 = -H$, $M_w = Wz/2$

(2) $\sum \int m_1^2 dz/EI = \int_0^H z^2 dz/EI + \int_0^{L/2} H^2 dz/2EI = H^2(4H+3L)/12EI$

$\sum \int m_1 M_w dz/EI = -\int_0^{L/2} WHz\, dz/4EI = -WHL^2/32EI$

$R_1 = -\sum \int m_1 M_w dz/EI \ / \ \sum \int m_1^2 dz/EI = 3WL^2/8H(4H+3L)$

(3) Determine the bending moments by superposition, i.e.
$M = M_w + m_1 R_1$

At B:
$M_w = 0$, $m_1 = -H$

Hence
$M_B = -HR_1 = -3WL^2/8(4H+3L)$

At C:
$M_w = WL/4$, $m_1 = -HR_1$

Hence
$M_C = WL/4 - 3WL^2/8(4H+3L) = WL(8H+3L)/8(4H+3L)$

7.10 Portal frames

Derive expressions for M_w and m_1 for the portal frames in Fig. 7.21, showing the limits of integration.

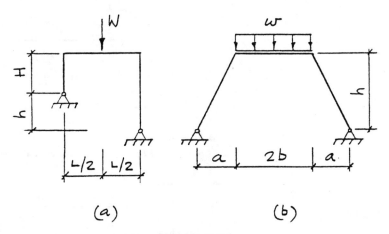

(a)

(b)

Figure 7.21

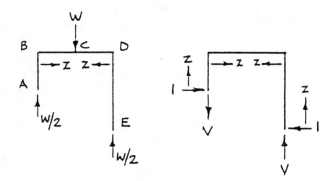

Figure 7.22

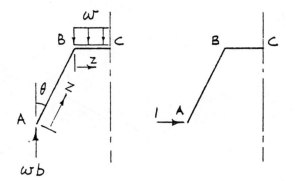

Figure 7.23

Solution (a) — Fig. 7.22

	M_w	m_1	Limits
AB	0	$-z$	0 to H
ED	0	$-z$	0 to $H+h$
BC	$Wz/2$	$-(Vz+H)$	0 to $L/2$
DC	$Wz/2$	$Vz-H-h$	0 to $L/2$

where $V = h/L$

Solution (b) − *Fig. 7.23* Symmetrical, so consider half the frame

	M_w	m_1	Limits
AB	$wbz\sin\theta$	$-z\cos\theta$	0 to $h\sec\theta$
BC	$wb(a+z) - wz^2/2$	$-h$	0 to b

where $\theta = \arctan(a/h)$

7.11 Two-pinned arch

A two-pinned parabolic arch has span L and height H and supports a central point load W. Derive an expression for the horizontal reaction at the supports assuming that the second moment of area at a horizontal distance x from the crown is given by $I = I_C\sec\alpha$, where I_C is the second moment of area at the crown and α is the angle the centre-line of the rib makes with the horizontal.

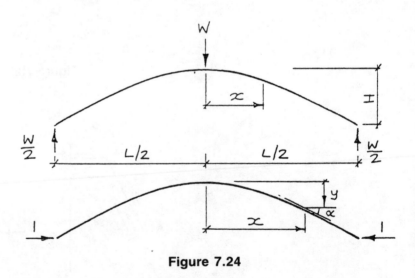

Figure 7.24

Solution − *Fig. 7.24* The assumption regarding the second moment of area enables the integration to be carried out. The result approximates to that for a uniform arch rib. A further simplification is to ignore the energy due to axial deformations.

(1) Let the horizontal reaction be R_1, the redundant restraint.

Then $M_w = W/2 \times (L/2 - x)$, $m_1 = -(H - y)$

The equation of the arch is $y = kx^2$. Putting $y = H$ and $x = L/2$,

$k = 4H/L^2$ and hence $y = 4Hx^2/L^2$
$m_1 M_w = WH/4 \times (4x^2/L - 8x^3/L^2 - L + 2x)$
$m_1^2 = H^2(16x^4/L^4 - 8x^2/L^2 + 1)$

(2) Denoting distance along the rib by s, the horizontal reaction is given by

$$R_1 = - \int m_1 M_w ds/EI \; / \int m_1^2 ds/EI$$

As it stands this expression cannot be integrated because $m_1 M_w$ and m_1^2 are in terms of x, the horizontal distance. However, if it is assumed that $I = I_C \sec\alpha$

then $I = I_C ds/dx$
i.e. $ds/EI = dx/EI_C$

As I_C is a constant the integration is now straightforward. Making use of symmetry only half the arch need be considered, thus

$$\int m_1 M_w ds/EI = WH/4EI_C \times \int_0^{L/2}(4x^2/L - 8x^3/L^2 - L + 2x)dx$$
$$= -5WHL^2/96EI_C$$

$$\int m_1^2 ds/EI = H^2/EI_C \times \int_0^{L/2}(16x^4/L^4 - 8x^2/L^2 + 1)dx = 4H^2L/15EI_C$$

Hence
$R_1 = 25WL/128H$

7.12 Thermal strain

Determine the horizontal reaction to a rectangular pinned-base portal frame of height H and span L due to a temperature rise of $t\,°C$. The coefficient of linear expansion is $\alpha/°C$. EI is constant.

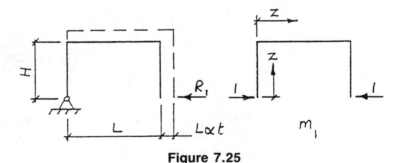

Figure 7.25

Solution − Fig. 7.25   Removal of the horizontal reaction R_1 allows the base to move laterally. The effect of the thermal expansion is then shown as the dotted line in the figure and an elastic horizontal deflection of $L\alpha t$ is required to restore the base to its original position. Now, considering R_1 as an external load, this deflection is given by

$$\delta = \Sigma \int m_1 M_w dz / EI = L\alpha t$$

where m_1 is the bending moment due to a unit force in the direction of R_1 and, as there are no other external loads,

$$M_w = m_1 R_1$$

Hence

$$R_1 = EIL\alpha t / \Sigma \int m_1^2 dz$$

Now m_1(columns) $= -z$, m_1(beam) $= -H$

Hence

$$\Sigma \int m_1^2 dz = 2 \int_0^H z^2 dz + \int_0^L H^2 dz = 2H^3/3 + H^2 L$$

and $R_1 = 3EIL\alpha t / H^2(2H/L + 3)$

7.13 Two-span beam by tabulated integrals

A uniform continuous beam of total length $2L$ is simply supported at the ends and at mid-span and carries a uniformly distributed load of intensity w. Find the value of the reactions, and the bending moment over the central support.

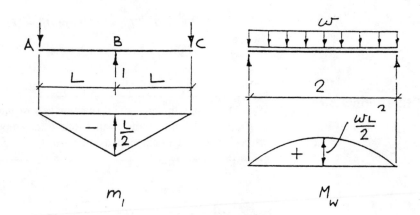

Figure 7.26

Solution — Fig. 7.26 (1) Let the central reaction be the redundant restraint R_1 and sketch the bending moment diagrams for m_1 and M_w, as shown.

(2) Since EI is constant

$$R_1 = -\Sigma \int m_1 M_w dz / \Sigma \int m_1^2 dz$$

(3) Evaluate the integral products using Table 6.1 and equation (6.11), i.e. $\int_0^L m_i m_j dz = \alpha L a c$

		α	L	a	c
$m_1 M_w$	$= \triangle \times \frown$	5/12	2L	$-L/2$	$wL^2/2$
m_1^2	$= \triangle \times \triangle$	1/3	2L	$-L/2$	$-L/2$

Hence

$$R_1 = [5/12 \times 2L \times (-L/2) \times wL^2/2]/[1/3 \times 2L \times (-L/2)] = 5wL/4$$

(4) Resolving vertically

$$R_A = R_B = wL - R_1/2 = 3wL/8$$

and $\quad M_B = R_A L - wL^2/2 = -wL^2/8$

7.14 Portal frame by tabulated integrals

Draw the shear force and bending moment diagrams for the pinned-base portal frame in Fig. 7.27. Young's modulus is constant and the second moments of area are as shown.

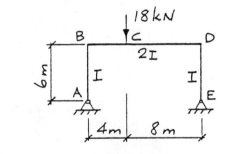

Figure 7.27

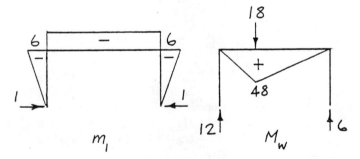

Figure 7.28

Solution — *Figs 7.28, 7.29* (1) Let the horizontal reaction at E be the redundant restraint R_1. Sketch the bending moment diagrams for m_1 and M_w, as shown.

(2) Since E is constant

$$R_1 = -\Sigma \int m_1 M_w dz/I \ / \ \Sigma \int m_1^2 dz/I$$

(3) Evaluate the integral products using kN, m units.

		α	L	a	c	EI
m_1M_w (AB, DE) $= 0$						
(BD) $= \square \times \triangle$		1/2	12	-6	48	2

$$\int m_1 M_w dz/EI = \Sigma \alpha L a c/EI = -864$$

		α	L	a	c	EI
m_1^2 (AB, DE) $= 2 \times \triangle \times \triangle$		1/3	6	-6	-6	1
(BD) $= \square \times \square$		1	12	-6	-6	2

$$\int m_1^2 dz/EI = \Sigma \alpha L a c/EI = 2 \times 72 + 216 = 360$$

Hence

$$R_1 = 864/360 = 2.4 \, \text{kN}$$

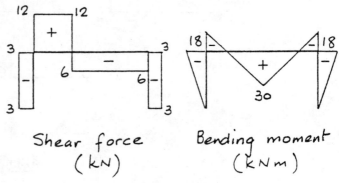

Shear force
(kN)

Bending moment
(kNm)

Figure 7.29

(4) Calculate shear forces and bending moments and draw the diagrams (Fig. 7.29).

$$S_{AB} = S_{ED} = -2.4 \, \text{kN}, \quad S_{BC} = 12 \, \text{kN}, \quad S_{CD} = 12 - 18 = -6 \, \text{kN}$$
$$M = M_w + m_1 R_1$$

Hence

$$M_A = 0, \quad M_B = M_D = -6 \times 2.4 = -14.4 \, \text{kNm},$$
$$M_C = 48 - 14.4 = 33.6 \, \text{kNm}$$

7.15 Pitched-roof portal frame by tabulated integrals

Determine the bending moments at the crown, eaves and supports of the pitched-roof frame in Fig. 7.30. Assume EI is constant.

Solution – Fig. 7.31 (1) Since the frame is symmetrical consider the left hand half only and select as redundancies the reactions at the crown from the right hand half, i.e. bending moment R_1, horizontal force R_2, and the vertical

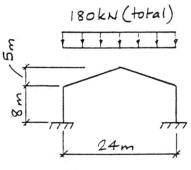

Figure 7.30

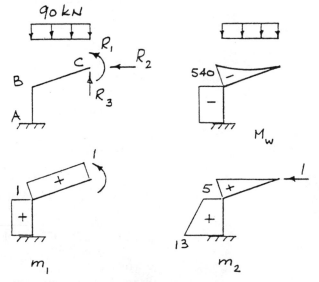

Figure 7.31

force R_3. Owing to symmetry the vertical reactions at the supports are equal to half the total load, so R_3 must be zero.

(2) Draw bending moment diagrams for M_w, m_1, and m_2 as shown. Note that the unit load for m_1 is a couple.

(3) Write two equations in terms of the redundancies R_1 and R_2 by putting $q = 1,2$ in equation 7.11 and multiplying through by EI, thus

$$R_1\Sigma\int m_1^2 dz + R_2\Sigma\int m_1 m_2 dz = -\Sigma\int m_1 M_w dz$$
$$R_1\Sigma\int m_2 m_1 dz + R_2\Sigma\int m_2^2 dz = -\Sigma\int m_2 M_w dz$$

(4) Evaluate the integrals, using kN, m units. The integrals within the summations are for the column AB and the rafter BD. Units are kN and m. The diagram of m_2 for AB is trapezoidal, for which there is no entry in Table 6.1. A value of α can be derived for trapezoidal diagrams, but a combination of a rectangle and a triangle can just as easily be used.

	α	L	a	c
m_1^2 (AB) $= (\square)^2$	1	8	1	1
(BC) $= (\square)^2$	1	13	1	1

$$\int m_1^2 dz = \Sigma \alpha Lac = 21$$

	α	L	a	c
$m_1 m_2$ (AB) $= \square \times \square$	1	8	1	5
$+ \square \times \triangleright$	1/2	8	1	8
(BC) $= \square \times \triangleright$	1/2	13	1	5

$$\int m_1 m_2 dz = \int m_2 m_1 dz = \Sigma \alpha Lac = 104.5$$

	α	L	a	c
m_2^2 (AB) $= (\square + \triangleright)^2$				
$= (\square)^2$	1	8	5	5
$+ 2(\square \times \triangleright)$	1/2	8	5	8
$+ (\triangleright)^2$	1/3	8	8	8
(BC) $= (\triangleright)^2$	1/3	13	5	5

$$\int m_2^2 dz = \Sigma \alpha Lac = 799$$

	α	L	a	c
$m_1 M_w$ (AB) $= \square \times \square$	1	8	1	-540
(BC) $= \square \times \triangleright$	1/3	13	1	-540

$$\int m_1 M_w dz = \Sigma \alpha Lac = -6.660\text{E}3$$

	α	L	a	c
$m_2 M_w$ (AB) $= \square \times \square$	1	8	5	-540
$+ \triangleright \times \square$	1/2	8	8	-540
(BC) $= \square \times \triangleright$	1/4	13	5	-540

$$\int m_2 M_w dz = \Sigma \alpha Lac = -47.66\text{E}3$$

Adding the results for AB and BC,

$$21.00R_1 + 104.5R_2 = 6.660\text{E}3$$
$$104.5R_1 + 799.0R_2 = 47.66\text{E}3$$

Note: With practice detailed setting out of the calculations as above becomes unnecessary (see next problem) except as an aid to checking.

Solving, $R_1 = 58.18$ kNm, $R_2 = 52.04$ kN.

(5) Determine the bending moments at the eaves and supports.

i.e. $M = M_w + m_1 R_1 + m_2 R_2$
Hence
$$M_A = -540 + 58.18 + 13 \times 52.04 = 194.7 \text{ kNm}$$
and $M_B = -540 + 58.18 + 5 \times 52.04 = -221.6 \text{ kNm}$
$$M_C = R_1 = 58.18 \text{ kNm}$$

7.16 Asymmetrical arch

A parabolic arch of span L and rise H has supports at the same level, one fixed and the other pinned, and carries a central point load W. Assuming that $I_x = I_C \sec\alpha$, determine the support reactions and the bending moments at the crown and the fixed support.

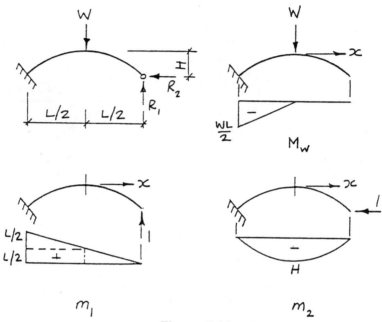

Figure 7.32

Solution – *Fig. 7.32* (1) Select the reactions at the pinned support as the redundant restraints R_1 (vertical) and R_2 (horizontal). The residual statically determinate system when both restraints are set to zero is then a cantilever and the bending moment diagrams M_w, m_1 and m_2 are as shown. Since the lever arm for R_2 is the height of the arch, m_2 is a parabola with the same shape as the arch rib.

(2) Use the assumption $I_x = I_C \sec\alpha$ to enable integration along the x axis, as in Example 7.11, and then divide through by the constant I_C to obtain the following equations:

$$R_1 \int m_1^2 dx + R_2 \int m_1 m_2 dx = -\int m_1 M_w dx$$
$$R_1 \int m_2 m_1 dx = R_2 \int m_2^2 dx = -\int m_2 M_w dx$$

Using $\int m_i m_j dx = \alpha Lac$ and referring to Table 6.1

$$R_1 \times 1/3 \times L \times L^2 + R_2 \times 1/3 \times L \times L \times (-H)$$
$$= -(1/2+1/3) \times L/2 \times L/2 \times (-WL/2)$$
$$R_1 \times 1/3 \times L \times (-H) \times L + R_2 \times 8/15 \times L \times H^2$$
$$= -1/4 \times L/2 \times (-H) \times (-WL/2)$$

Solving, $R_1 = 25W/48$, $R_2 = 5WL/24H$

The trussed girder in Fig. 7.33 consists of a beam braced by a prop at mid-span and two ties. Joints at the ends of the prop and the ties may be assumed to be pinned and there is no lack of fit. The members, which are all of the same material, have the following section properties. Second moment of area of beam 40E6 mm^4; cross sectional areas: beam 3200 mm^2, prop 3200 mm^2, ties 500 mm^2. Taking into account the energy due to both bending and axial force, determine the axial forces in all the members and the bending moment in the beam at mid-span when it carries a uniformly distributed load of 20 kN/m.

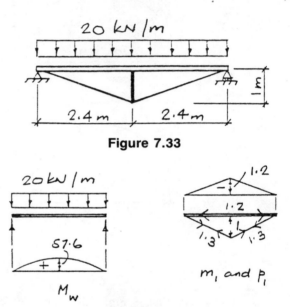

Figure 7.33

Figure 7.34

Solution — Fig. 7.34 (1) There is only one redundancy. Let this be R_1, the force in the prop. Replacement of the prop by a unit force produces axial forces p_1 in all the members and bending moments m_1 in the beam, as shown. The external load produces axial forces P_w and bending moments M_w. However, since the external load is supported entirely by the beam when R_1 is set to zero, P_w is the zero in all the members and M_w exists only in the beam.

(2) Since the total energy is the simple sum of the energies due to axial forces and bending moments (see Chapter 1),

$$\Sigma(P_w + p_1R_1)p_1L/AE + \Sigma\int(M_w + m_1R_1)m_1 \; dz/EI = 0$$

Setting all $P_w = 0$ and transposing,

$$R_1 = -\Sigma\int m_1 M_w dz/I \; / \; (\Sigma p_1^2 L/A + \Sigma\int m_1^2 dz/I)$$

Working in kN and m units, first for the axial forces:

$$\sum p_1^2 L/A = 2 \times (1.3)^2 \times 2.6/500E\text{-}6 + 1^2 \times 1/3200E\text{-}6$$
$$+ (1.2)^2 \times 4.8/3200E\text{-}6 = 20.05E3$$

Integrating the bending moments by means of Table 6.1.

$$L = 4.8, I = 40E\text{-}6, a = -4.8/4 = -1.2, c = 20 \times 4.8^2/8 = 57.6$$
$$\int m_1^2 dz/I = \alpha L a^2/I = 1/3 - 4.8 \times (-1.2)^2/40E\text{-}6 = 57.6E3$$
$$\int m_1 M_w dz/I = \alpha L a c/I = 5/12 \times 4.8 \times (-1.2) \times 57.6/40E\text{-}6$$
$$= -3.456E6$$

Hence
$$R_1 = 3.456E6/(20.5E3 + 57.6E3) = 44.51 \text{ kN}$$

(3) Axial forces: $P = p_1 R_1$

Ties: $P = 1.3R_1 = 57.86 \text{ kN}$
Beam: $P = -1.2R_1 = -53.41 \text{ kN}$
Prop: $P = -R_1 = -44.51 \text{ kN}$

Bending moments: $M = M_w + m_1 R_1$

$$M_B = 57.6 - 1.2 \times 44.51 = 4.19 \text{ kNm}$$

7.18 Deflection of continuous beam

> A simply supported continuous beam carries a uniform load of 24 kN/m over two equal spans of 6 m. $EI = 13E3$ kNm². Determine the central deflection in each span, using the results of Example 7.13.

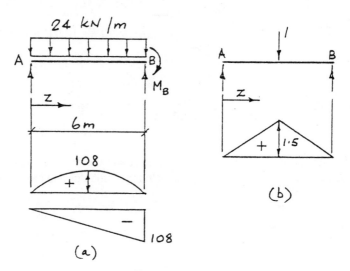

Figure 7.35

Solution — Fig. 7.35 (1) As the beam is symmetrical the deflections in the spans are equal, so consider the span AB as a free body, in which the central deflection is given by

$$\delta = \Sigma \int mM \, dz/EI$$

where M is the bending moment due to the external loading in Fig. 7.35(a), including the redundant moment M_B which was determined in Example 7.13, and m is the bending moment due to a unit vertical load at the centre of AB, as in Fig. 7.35(b).

(2) If the problem is solved by direct integration the expressions for M and m are as follows.

$$R_A = 24 \times 3 - M_B/6$$
$$M = R_A z - wz^2/2$$
$$m = z/2 \text{ for } z \le 3$$
$$m = (6-z)/2 \text{ for } z > 3$$

(3) Alternatively, if Table 6.1 is used the bending moment M may be considered as the sum of the two diagrams in Fig. 7.35(a), i.e. the free bending moment due to the distributed load, with maximum value

$$M_{max} = wL^2/8 = 24 \times 6^2/8 = 108 \text{ kNm}$$

and the fixing moment diagram, with maximum value

$$M_B = -wL^2/8 \text{ (from Example 7.13)} = -108 \text{ kNm}$$

The maximum value of m is $L/4$, i.e. 1.5 kNm.
Hence

mM	$= \triangle \times \bigtriangleup$	1/3	6	1.5	108
	$- \triangle \times \measuredangle$	1/4	6	1.5	−108

$$\delta = \Sigma \alpha Lac/EI = 162/13E3 = 12.46E\text{-}3 \text{ m}$$

Problems

Pin-jointed frameworks

1 Determine the member forces for the truss in Fig. 7.36, assuming that Young's modulus and the cross sections of the members are constant.

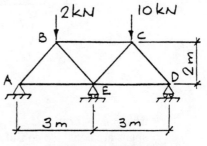

Figure 7.36

Solution

Let R_1 be the reaction at E. Units: kN, m.

Member	p_1	P_w	L	$P=P_w+p_1R_1$
AB	0.625	−5.0	2.5	−0.76
BC	0.750	−4.5	3.0	0.59
CD	0.625	−10.0	2.5	−5.76
CE	−0.625	−2.5	2.5	−6.74
BE	−0.625	2.5	2.5	−1.74
AE	−0.375	3.0	3.0	0.45
DE	−0.375	6.0	3.0	3.45

$$R_1 = -\Sigma p_1 P_w L / \Sigma p_1^2 L = 43.69/6.438 = 6.79 \text{ kN}$$

2 Find the member forces for the truss in Fig. 7.37. All members have $AE = 20\,000$ kN. Member BE was 2.5 mm short on assembly.

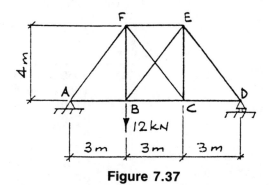

Figure 7.37

Solution

Let R_1 be the force in member BE. Units: kN, m.

Member	p_1	P_w	L	$P=P_w+r_1R_1$
AB	0.0	6	3	6.00
BC	−0.6	6	3	1.43
CD	0.0	3	3	3.00
DE	0.0	−5	5	−5.00
EF	−0.6	−3	3	−7.57
FA	0.0	−10	5	−10.00
FB	−0.8	12	4	5.91
EC	−0.8	4	4	−2.09
FC	1.0	−5	5	2.62
BE	1.0	0	5	7.62

$$R_1 = (-\lambda AE + \Sigma r_1 P_w L)/\Sigma r_1^2 L = (50+81.6)/17.28 = 7.616\ \text{kN}$$

3 Find the member forces in the truss shown in Fig. 7.38. EA is constant and there is no lack of fit or settlement.

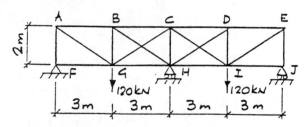

Figure 7.38

Solution
Let R_1 be the reaction at H, and R_2 the force in members CG and CI. Units: kN, m.

Member	r_1	r_2	P_w	L	$P=P_w+r_1R_1+r_2R_2$
AB	0.750	0.000	−180.0	3	−54.00
BC	1.500	−0.832	−180.0	3	11.96
FG	0.000	0.000	0.0	3	0.00
GH	−0.750	−0.832	180.0	3	−6.04
AF	0.500	0.000	−120.0	2	−36.00
BG	0.500	−0.555	0.0	2	43.95
AG	−0.901	0.000	216.3	$\sqrt{13}$	64.93
BH	−0.901	1.000	0.0	$\sqrt{13}$	−79.21
CG	0.000	1.000	0.0	$\sqrt{13}$	72.16
CH	0.000	−1.109	0.0	2	−80.03

$$R_1\Sigma r_1^2 L + R_2 \Sigma r_1 r_2 L = \Sigma r_1 P_w L = 0$$
$$R_1 \Sigma r_2 r_1 L + R_1 \Sigma r_2^2 L + \Sigma r_2 P_w L = 0$$
In summations repeat for all members except CH. Hence
$$33.96 R_1 - 11.35 R_2 - 4.885 = 0$$
$$-11.35 R_1 + 26.42 R_2 + 0.000 = 0$$
Solving, $R_1 = 168.0\ \text{kN}$, $R_2 = 72.16\ \text{kN}$

4 Find the forces in all the members and the deflection at joint E of the truss in Fig. 7.39. $EA = 4000\ \text{kN}$ for all members.

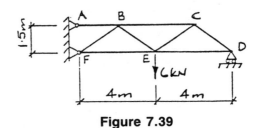

Figure 7.39

Solution

Let R_1 be the reaction at D. Find the member forces. Then for the deflection apply a unit force at E to obtain influence coefficients p. Units: kN, m.

Member	r_1	p	$P_w=6p$	L	$P=P_w+r_1R_1$
AB	−16/3	8/3	16	2	4.294
BC	−8/3	0	0	4	−5.853
DE	4/3	0	0	4	2.926
EF	12/3	−4/3	−8	4	0.779
CD	−5/3	0	0	5	−3.658
BE	−5/3	5/3	10	5	6.342
CE	5/3	0	0	5	3.658
BF	5/3	−5/3	−10	5	−6.342

$$R_1 = -\Sigma r_1 P_w L / \Sigma r_1^2 L = 465.3/212.0 = 2.195 \text{ kN}$$
$$\delta = \Sigma pPL/EA = 124.4/4000 = 0.031 \text{ m, i.e. } 31 \text{ mm}$$

Flexural structures

In the solutions to the following problems the first of the two letters denoting the ends of a member is taken as the origin of z. The limits of integration (except in Problem 9) are therefore zero and the length of the member.

In most cases integration may be performed either directly or by means of Table 6.1. The former method has been used in Problems 5 to 9. It is suggested that (with the exception of Problem 9) the alternative method be tried on a second reading.

5 Find the reactions at the feet of the portal frame in Fig. 7.40 and determine the bending moments at the ends and mid-span of the beam BC. *EI* is constant.

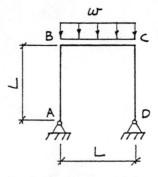

Figure 7.40

Solution

Vertical reactions $= wL/2$, by symmetry. Let R_1 be the horizontal reaction.

AB and DC:	$m_1 = -z$	$M_w = 0$	
BC:	$m_1 = -L$	$M_w = wLz/2 - wz^2/2$	

$$R_1 = -\sum \int m_1 M_w dz / \sum \int m_1^2 dz = wL/20$$
$$M_B = M_C = -wL^2/20 \quad M_{BC} = 3wL^2/40$$

6 The frame in Fig. 7.41 has constant *EI* and is pinned at supports A and E. Find the bending moments at B, C and D.

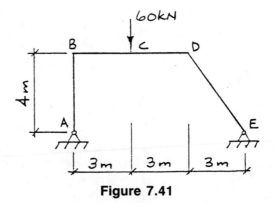

Figure 7.41

Solution

Let R_1 be the horizontal reaction. Vertical reactions when simply supported: $V_A = 40\,\text{kN}$, $V_E = 20\,\text{kN}$.

	AB	BC	CD	ED
m_1	$-z$	-4	-4	$-0.8z$
M_w	0	$40z$	$20(6-z)$	$20 \times 0.6z = 12z$

$$R_1 = -\Sigma \int m_1 M_w dz / \Sigma \int m_1^2 dz = 2200/144 = 15.28 \text{ kN}$$
$$M_B = -61.1 \text{ kNm} \qquad M_C = 58.9 \text{ kNm} \qquad M_D \ -1.1 \text{ kNm}$$

7 The portal frame of Fig. 7.42 has pinned supports at A and D. Assuming a constant EI, determine:
(a) the shear forces at the ends of the beam BC
(b) the maximum bending moment in the beam.

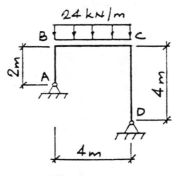

Figure 7.42

Solution
Let R_1 be the horizontal reaction (inwards) at A and D: then treat as simply supported. Vertical reactions: (a) to external load, $V_A = V_D = 48 \text{ kN}$; (b) to unit load, $V_A = 1/2 \text{ kN}$ (downwards), $V_D = 1/2 \text{ kN}$ (upwards).

	AB	BC	DC
m_1	$-z$	$-(2+z/2)$	$-z$
M_w	0	$48z - 12z^2$	0

$$R_1 = -\Sigma \int m_1 M_w dz / \Sigma \int m_1^2 dz = 384/61.33 = 6.26 \text{ kN}$$
(a) $S_{BC} = 48 - R_1/2 = 44.87 \text{ kN}$
$\quad S_{CB} = -(48 + R_1/2) = -51.13 \text{ kN}$
(b) At point of zero shear force,
$\quad z = 4 \times 44.87/(44.87 + 51.13) = 1.87 \text{ kN}$
$\quad M = M_w + m_1 R_1$
$\quad$ Hence $M_{max} = 48 \times 1.87 - 12 \times 1.87^2 - (2 + 1.87/2) \times 6.26$
$\qquad\qquad\qquad = 29.42 \text{ kNm}$

8 Determine the bending moments at the supports and at mid-span of the built-in beam in Fig. 7.43. EI is constant.

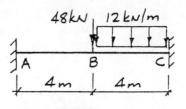

Figure 7.43

Solution

Let R_1 and R_2 be the bending moments at A and C; then treat as simply supported. Assume positive convention, i.e. R_1 clockwise, R_2 anticlockwise.

AB:	$m_1 = 1-z/8$	$m_2 = z/8$	$M_w = 36z$
CB:	$m_1 = z/8$	$m_2 = 1-z/8$	$M_w = 60z-z^2$

$R_1\Sigma\int m_1^2 dz + R_2\Sigma\int m_1 m_2 dz + \Sigma\int m_1 M_w dz = 0$
$R_1\Sigma\int m_2 m_1 dz + R_2\Sigma\int m_2^2 dz + \Sigma\int m_2 M_w dz = 0$
i.e. $2.667R_1 + 1.333R_2 + 304 = 0$
$\quad\quad 1.333R_1 + 2.667R_2 + 336 = 0$
Solving, $R_1 = -68.0$ kNm, $R_2 = -92$ kNm; and hence
$M_B = 64$ kNm

9 A two-pinned parabolic arch has a span of 120 m and carries a vertical point load of 480 kN at the crown. The supports are 10 m and 40 m below the level of the crown. Given that $I = I_C \sec\alpha$, where I_C is the second moment of area at the crown and α is the inclination of the axis of the rib from a horizontal line through the crown, determine the reactions at the supports and the bending moment at the crown.

Solution — Fig. 7.44 (dimensions in m, loads in kN)

(1) Locate the crown and establish the equation of the arch (see Volume 1 of *Solving Problems in Structures*)
$x_A = 120/[1+\sqrt{(40/10)}] = 40$ m $\quad y = x^2/160$

(2) Let R_1 be the horizontal reaction; then treat as simply supported.
CA: $M_w = 320(40-x)$ $\quad m_1 = x^2/160 + x/4 - 20$ limits 0 to 40
CB: $M_w = 160(80-x)$ $\quad m_1 = x^2/160 - x/4 - 20$ limits 0 to 80

(3) $R_1 = -\Sigma\int m_1 M_w dx / \Sigma\int m_1^2 dx = 1000(3.84+10.24)/(25.6+6.8)$
$\quad = 434.6$ kN

Hence
$V_A = 211.4$ kN, $V_B = 268.7$ kN, $M_C = 4110$ kNm

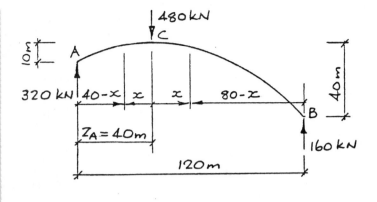

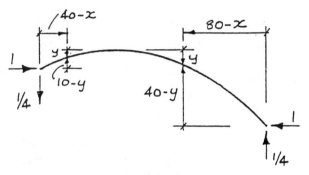

Figure 7.44

10 Calculate the principal bending moments for the frame in Fig.7.45. EI is constant.

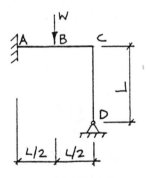

Figure 7.45

Solution – Fig. 7.46

    (1) Let the reactions at D be the redundancies and treat as a cantilever. Sketch the free bending moment diagrams for M_w, m_1, and m_2

    (2) $R_1\Sigma\int m_1^2 dz + R_2\Sigma\int m_1 m_2 dz + \Sigma\int m_1 M_w dz = 0$

            $R_1\Sigma\int m_2 m_1 dz + R_2\Sigma\int m_2^2 dz + \Sigma\int m_2 M_w dz = 0$

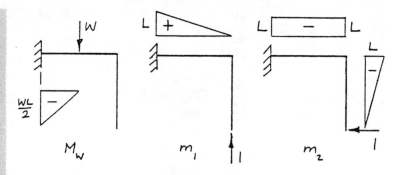

Figure 7.46

Using Table 6.1
$$R_1L^3/3 - R_2L^3/2 - 5WL^3/48 = 0$$
$$-R_1L^3/2 + R_2L^3 \times 4/3 + WL^3/8 = 0$$
Solving, $R_1 = 11W/28$, $R_2 = 3W/56$
(3) $M = M_w + m_1R_1 + m_2R_2$
Hence
$$M_A = -9WL/56, \quad M_B = WL/7, \quad M_C = -3WL/56, \quad M_D = 0$$

11 A vertical cable mast 32 m tall is built in at the base and is prestressed by a guy which was 20 mm short on assembly. The guy is stretched between a point at mid-height on the mast and an anchor at ground level, 12 m from the base. The cable exerts a force of 1 kN at the top of the mast, in the same plane as the guy. If EI of the mast is 40 000 kNm^2 and EA of the guy is 20 000 kN, determine (a) the force in the guy and (b) the horizontal deflections at the top and mid-height of the mast.

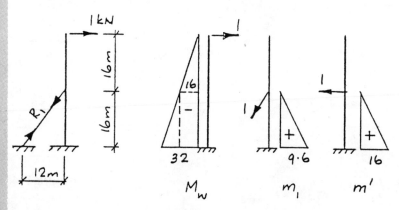

Figure 7.47

Solution — Fig. 7.47
(a) Let R_1 be the force in the guy. Consider bending only in the mast and tension only in the guy.

$$R_1 = -[(r_1\lambda + r_1 P_w L/EA)_{guy} + (\int m_1 M_w dz/EI)_{mast}]/$$
$$[(r_1^2 L/EA)_{guy} + (\int m_1^2 dz/EI)_{mast}]$$

where $r_1 = 1$, $P_w = 0$, $\lambda = -0.02$ m, m_1 and M_w as shown. Hence, using Table 6.1,

$$R_1 = -[-0.02 + 0 - 0.0512]/[0.001 + 0.01229] = 5.357 \text{ kN}$$

(b) For deflections apply unit loads at the top and mid-height to give m and m', where $m = M_w$ and m' is as shown. Hence, using Table 6.1,

$$\delta_{top} = \int (M_w + m_1 R_1) m \ dz/EI = -0.00121 \text{ m (i.e. to left)}$$
$$\delta_{mid} = \int (M_w + m_1 R_1) m' \ dz/EI = 0.0243 \text{ m (i.e. to left)}$$

12 The portal frame in Fig. 7.48 is built-in at A and pinned at E. Assuming constant EI, determine the reactions at E and the bending moments at A,B,C, and D. If $EI = 95\,000$ kNm^2 calculate the vertical deflection at C.

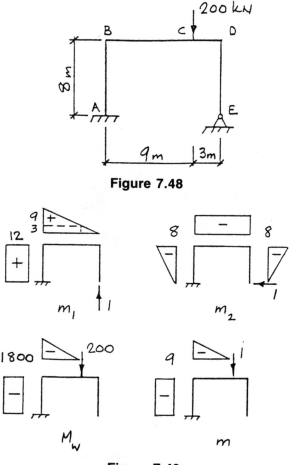

Figure 7.48

Figure 7.49

Solution − Fig. 7.49

(1) Let R_1 and R_2 be the vertical and horizontal reactions at E.

Then $R_1\Sigma\int m_1^2 dz + R_2\Sigma\int m_1 m_2 dz + \Sigma\int m_1 M_w dz = 0$

$R_1\Sigma\int m_2 m_1 dz + R_2\Sigma\int m_2^2 dz + \Sigma\int m_2 M_w dz = 0$

Using Table 6.1

$1.728 R_1 = 0.960 R_2 - 245.7 = 0$

$-0.960 R_1 + 1.109 R_2 + 122.4 = 0$

Solving $R_1 = 155.8$ kN, $R_2 = 24.49$ kN

(2) $\delta = \Sigma\int (M_w + m_1 R_1 + m_2 R_2) m\, dz / EI = 18.8\text{E-}3$ m,
i.e. 18.8 mm (using Table 6.1)

8

Influence Lines

Introduction

An *influence coefficient* is the value of some quantity such as reactive force, internal stress resultant, or displacement, produced by a unit load acting at some point on the structure. An *influence line* is a diagram showing how a particular influence coefficient varies as the unit load travels along one or more members of the structure. The difference between an influence line for a stress resultant and a stress resultant diagram should be clearly understood. *An influence line shows the value of the stress resultant at a particular point on the beam as the load moves along it, whereas a stress resultant diagram shows how, for a stationary load, the stress resultant varies as the point of interest moves along the beam.*

The virtue of an influence line is that it enables the value of the quantity to be determined for a variety of load-cases without re-analysing the structure. Influence lines are chiefly used in design and are especially useful in the design of long, bridge-type structures subjected to travelling loads. In diagrams a moving load is usually shown enclosed in a circle, symbolising a wheel or an axle.

Influence coefficients of forces and moments in statically determinate structures are linear functions of the position of the unit load. The corresponding influence lines are therefore composed entirely of straight lines. Influence lines for displacements and for all quantities related to hyperstatic structures are curved.

Statically determinate beams

The influence lines for the reactions of a simply supported beam are straight lines with ordinates varying from unity when the unit load is over the support concerned, to zero when it is over the far support, as shown in Fig. 8.1(a). When the load is at some point X the reactions are given by the ordinates r_A and r_B under the load, as shown. By simple proportion, $r_A = (L-x)/L$ and $r_B = x/L$.

The influence lines for stress resultants are shown by the full lines in Fig. 8.1(b). Thus the shear force and bending moment at point Z when the unit load is at point X are given by the ordinates s_Z and m_Z in Fig. 8.1(b). These diagrams can be constructed directly by placing the unit load at Z and

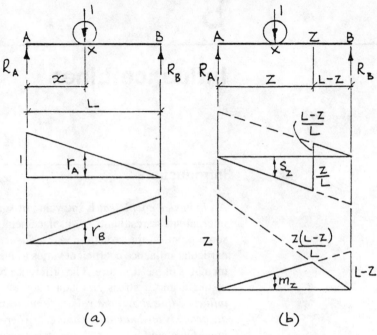

Figure 8.1

determining the stress resultants by statics, which gives the values shown in the figure. The diagram is then completed by joining to zero at the ends by straight lines. Alternatively the reaction diagrams can be used as follows. The shear force s_Z is given by $-r_B$ when the load is to the left of Z and by r_A when it is to the right. The diagram for r_B must therefore be inverted, as shown. The dotted lines show the parts of the reaction diagrams which are not relevant because of the position of the load. Similarly the bending moment m_Z is given by $r_B(L-z)$ when the load is to the left of Z and by $r_A z$ when it is to the right. Hence the diagram can be constructed by erecting ordinates z and $L-z$ at A and B, as shown.

The effect of a point load W in any position is obtained from the influence line by multiplying the ordinate under the load by the value of the load. For example: $R_A = Wr_A$, $M_Z = Wm_Z$, and so on. It follows that the effect of a train of loads is the sum of the products of each load and its corresponding ordinate. For example, in Fig. 8.2(a) the shear force at Z is given by

$$S_Z = -W_1 s_1 + W_2 s_2 + W_3 s_3 \tag{8.1}$$

The effect of a distributed load is obtained by integrating the product of the load intensity and the corresponding ordinate along the length of the load. For example, in Fig. 8.2(b) the bending moment at Z is given by

$$M_Z = \int_0^c w m_Z \mathrm{d}x \tag{8.2}$$

When w is constant this is simply w times the area of the influence diagram under the load.

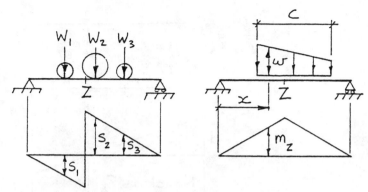

Figure 8.2

Statically determinate trusses

With modern methods of bridge construction there is little or no practical requirement for influence lines for trusses. However, they are still included in some syllabuses. It is usual for the load to be carried by a deck suspended between a pair of trusses and connected to the joints of either the bottom or top chord. A load lying between two joints is assumed to be distributed between them in the same way as the reactions of a simply supported beam.

The influence lines for reactions are the same as for beams. For member forces it is usually most convenient to use the *Method of Sections* (see Volume 1, *Solving Problems in Structures*) as the basis for analysis. The procedure is explained in Example 8.4.

Worked examples

8.1 Long distributed load

A beam ABC is simply supported at A and B, with AB 8 m and BC 4 m. Sketch the influence lines for the reactions at A and B, and the shear force and bending moment at a point D midway between A and B. Use the influence lines to determine the maximum positive and negative bending moments, and the maximum shear force, irrespective of sign, when a uniformly distributed load longer than AB and of intensity 20 kN/m crosses the beam.

Solution − *Fig. 8.3*   (1) Reactions: Unit load at A, $R_A = 1$, $R_B = 0$. Load at B, $R_A = 0$, $R_B = 1$. Complete the diagrams by joining the ordinates at A and B and producing the lines to C, as shown.

(2) Shear force: Load at A or B, $S_D = 0$. Load at D, $S_D = \mp 0.5$

(3) Bending moment: Load at A or B, $M_D = 0$. Load at D, $M_D = 1 \times 8/4 = 2$

(4) Complete the diagrams for shear force and bending moment by joining the ordinates at A, D and B and producing the lines to C, as shown.

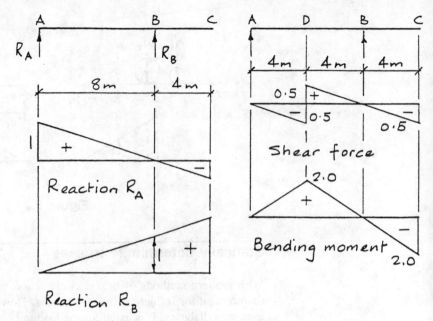

Figure 8.3

(5) Distributed load: Maximum positive bending moment occurs when the load covers the positive part of the diagram, i.e. the whole of AB.

$$M_D = w \times \text{area of diagram under AB} = 20 \times 8 \times 2/2 = 160 \, \text{kNm}$$

Similarly, to obtain the maximum negative bending moment BC must be fully loaded, giving

$$M_D = 20 \times 4 \times (-2)/2 = -80 \, \text{kNm}$$

In both the above cases, since the load is continuous and longer than AB, part of it must be assumed either not to have reached the beam or to have moved off it.

For the maximum shear force, again part of the load is off the beam, covering either AD or BC, both of which give a negative value. As the load is longer than AB there can be no positive shear force.

$$S_D = 20 \times 4 \times (-0.5)/2 = -20 \, \text{kN}$$

8.2 Load train

> The bridge beam in Fig. 8.4 is simply supported at A,B,E, and F, and hinged at C and D. Sketch the influence lines for the bending moment at B and the shear force at A. Use the diagrams to determine the maximum values when the train of loads shown crosses the bridge from left to right.

Solution – Fig. 8.5 (1) Bending moment at B: Unit load at A,B,D,E, or F: $M_B = 0$. Load at C: $M_B = -20 \, \text{kNm}$.

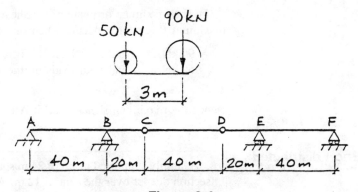

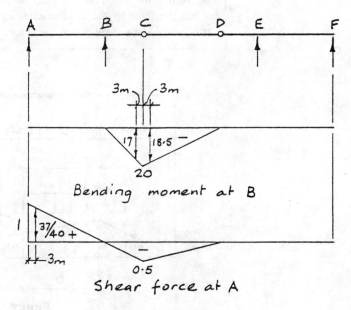

Figure 8.4

Figure 8.5

(2) Shear force at A: S_A = reaction at A. Load at B,D,E, or F: $S_A = 0$. Load at A: $S_A = 1$. Load at C: $S_A = -0.5$.

(3) It can be shown that the maximum bending moment occurs when one of the loads in the train is over the maximum ordinate of the influence line, i.e. at point C. It is therefore necessary to calculate the ordinates 3 m on either side of C. By simple proportion these are 17 and 18.5, as shown. The maximum bending moment is then either:

$$-(50 \times 17 + 90 \times 20) = -2650 \text{ kNm} \quad \text{with the 90 kN load at C, or}$$
$$-(50 \times 20 + 90 \times 18.5) = -2665 \text{ kNm} \quad \text{with the 50 kN load at C}$$

Hence the maximum bending moment at B is -2665 kNm, when the 50 kN load is at C.

(4) The maximum ordinate of the shear force influence line is at A, so the maximum shear force occurs when one of the loads is at A, i.e. either:

90 kN at A and 50 kN off the beam: $S = 90 \times 1 = 90$ kN

50 kN at A and both loads on the beam: $S = 50 \times 1 + 90 \times 37/40$
$= 133$ kN

Hence the maximum shear force at A is 133 kN, with the 50 kN load at A.

8.3 Short distributed load

A simply supported beam of 24 m span is strengthened by making the section deeper over the central 12 m. A uniformly distributed load 8 m long travels across the beam. By means of influence lines or otherwise prove that the maximum bending moment in the smaller section occurs when the front of the load reaches mid-span. Determine the value of this bending moment when the intensity of the load is 36 kN/m.

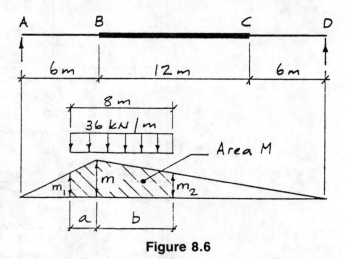

Figure 8.6

Solution − *Fig. 8.6* (1) When a unit load crosses the beam, the maximum bending moment in the smaller section occurs when the load is at the junction of the two sections, say at B. The influence line is then as shown.

(2) The maximum bending moment due to the distributed load occurs when the area M is maximal.

$$M = [a(m+m_1) + b(m+m_2)]/2$$

where $m_1 = m(6-a)/6$, $m_2 = m(18-b)/18$ by simple proportion. Putting the unit load at B,

$$m = 6 \times 18/24 = 4.5$$

Substituting for m_1 and m_2 and putting $a = 8-b$

$$M = 4.5(24 + 12b - b^2)/9 \qquad (1)$$

For this to be a maximum, $dM/db = 0$. Hence $b = 6$, i.e. the front of the load is at mid-span. (QED)

Note: It can be shown in general that the maximum bending moment at any point on a simply supported beam, due to a short uniformly distributed load, occurs when the point divides the load in the same proportion as it divides the span. Thus in this problem $b/8 = 18/24$, i.e. $b = 6$.

It can also be shown that this condition occurs when the ordinates m_1 and m_2 are equal. Substituting for a and b in this problem gives $m_1 = m_2 = 2m/3 = 3$.

(3) Multiplying area M by the intensity of the distributed load and putting $b = 6$ in equation (1),

$$M_{max} = 36M = 36 \times 30 = 1080 \, kNm \qquad (2)$$

8.4 Members of a simple truss

Sketch the influence line for the member forces X, Y, and Z of the truss in Fig. 8.7, assuming that the load travels along the bottom chord. Determine the maximum tensile and compressive values of Y due to a uniformly distributed load of intensity w, of any length, in the worst position.

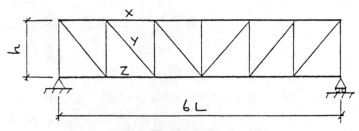

Figure 8.7

Solution − Fig. 8.8 (1) Cut the truss as shown and treat the parts of the frame on either side of the cut as free bodies in equilibrium under the action of external loads and the forces in the cut members (*Method of Sections*). Assume initially that all member forces are tensile. When the unit load is to the left of B member forces are determined by considering the right hand part of the frame and, since the only external load acting on this part is the reaction R_B, the influence lines to the left of B are linear functions of R_B. Similarly, when the load is to the right of C the left hand part of the frame is considered and the influence lines are linear functions of R_A. When the load is between B and C it is shared between the two joints, i.e. between the two parts. Therefore, as the member forces are zero when the load is over either support, it is only necessary to find their values when the load is at B and C to define the influence lines.

Load at B: $\quad R_B = 1/6$
Load at C: $\quad R_A = 2/3$

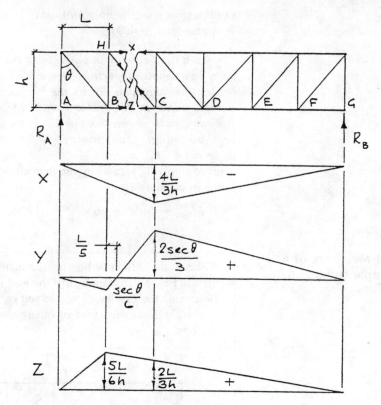

Figure 8.8

(2) Member force X: Take moments about C.

Load at B: $X = -R_B \times 4L/h = -2L/3h$
Load at C: $X = -R_A \times 2L/h = -4L/3h$

It can be seen from these results that there is no discontinuity at B because when the load is between B and C the portion of it acting at C exerts no moment about C. The first equation therefore holds between A and C.

(3) Member force Y: Resolve forces vertically.

Load at B: $Y = -R_B \sec\theta = -(\sec\theta)/6$
Load at C: $Y = R_A \sec\theta = 2(\sec\theta)/3$

where $\theta = \arctan(L/h)$.
Note: The sign of the member force changes as the load crosses from B to C.

(4) Member force Z: Take moments about H.

Load at B: $Z = R_B \times 5L/h = 5L/6h$
Load at C: $Z = R_A \times L/h = 2L/3h$

There is no discontinuity at C in this case, because when the load is between B and C the portion of it acting at B has no moment about H and the second equation therefore holds between B and G.

(5) The maximum values of X and Z occur when the distributed load occupies the whole span, thus

$$X_{max} = -4L/3h \times 6L/2 \times w = -4wL^2/h$$
$$Z_{max} = 5L/6h \times 6L/2 \times w = 5wL^2/2h$$

The maximum positive and negative values of Y occur when the load fills the span to the right and left respectivley of a point distant $L/5$ from B along BC. Hence

$$Y_{max}(+\text{ve}) = 2(\sec\theta)/3 \times 12L/5 \times w = 8wL/5 \times \sec\theta$$
$$Y_{max}(-\text{ve}) = -(\sec\theta)/6 \times 3L/5 \times w = -wL/10 \times \sec\theta$$

The reciprocal theorems

The derivation of influence lines for displacements and for hyperstatic structures is simplified by application of the reciprocal theorems of Maxwell and Betti.

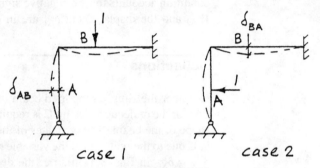

Figure 8.9

Consider the structure in Fig. 8.9, which is subjected to two different load-cases.

(a) A vertical unit load at B which produces a horizontal deflection δ_{AB} at A.

(b) A horizontal unit load at A which produces a vertical deflection δ_{BA} at B.

Maxwell's reciprocal theorem states that

$$\delta_{AB} = \delta_{BA} \tag{8.3}$$

The conditions are that:

(a) the structure must be linear and elastic

(b) the displacement in case 1 must be vectorially equivalent to the unit load in case 2, and vice versa.

Betti's reciprocal theorem is an extension of that of Maxwell. Consider, for example, the two load-cases on the structure in Fig. 8.10. The loads in case 1 produce displacements δ_{A1}, δ_{B1}, δ_{E1} in the lines of action of the loads in case 2, namely W_{A2}, W_{B2}, W_{E2}. Similarly the displacements θ_{C2}, δ_{D2}, δ_{F2} are produced by the loads in case 2, in the lines of action of the loads M_{C1}, W_{D1}, W_{F1} in case 1. Betti's theorem states that:

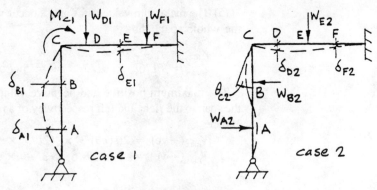

Figure 8.10

$$M_{C1}\theta_{C2} + W_{D1}\delta_{D2} + W_{F1}\delta_{F2} = -W_{A2}\delta_{A1} + W_{B2}\delta_{B1} + W_{E2}\delta_{E1} \qquad (8.4)$$

The conditions are the same as for Maxwell's reciprocal theorem. The second condition accounts for the negative sign in the above expression since the load W_{A2} and the displacement δ_{A1} are in opposite directions.

Deflections

Consider the simply supported beam in Fig. 8.11(a), for which the influence line for the deflection at point C is required. The dotted line shows the deflected shape of the beam. The ordinates of the influence line are δ_{CX}, the deflection at C due to the unit load at the variable position X. But, according to Maxwell, $\delta_{CX} = \delta_{XC}$ in Fig. 8.11(b), i.e. the deflection at X due to a unit load at C. The influence line for the deflection at C is therefore the deflected shape of the beam due to a stationary unit load at C, as shown by the dotted line in Fig. 8.11(b). It is usually easier to determine the influence line indirectly for a stationary load, e.g. by Macaulay's method, than directly by finding the deflections at a particular point due to a load whose position is variable.

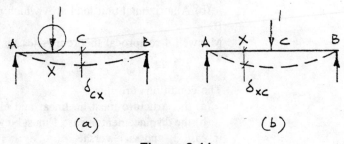

Figure 8.11

Hyperstatic structures

For hyperstatic structures the influence lines for the redundancies must first be obtained. From these the influence lines for other quantities may then be derived by statics. When there is only a single redundancy the equation of the influence line can be derived either directly by any of the standard methods, or indirectly, by the use of Maxwell's reciprocal theorem, as follows.

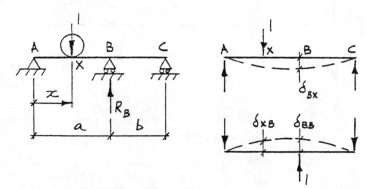

Figure 8.12

Suppose that the influence line for the reaction R_B to the two-span continuous beam in Fig. 8.12 is required. First remove R_B and let the deflection at B produced by a unit load at X be δ_{BX}. Next let the deflections produced by a unit load in place of R_B be δ_{BB} at B and δ_{XB} at X. Then, for compatibility,

$$\delta_{BX} - R_B \delta_{BB} = 0$$

But, by Maxwell's reciprocal theorem, $\delta_{BX} = \delta_{XB}$,
Hence
$$R_B = \delta_{XB}/\delta_{BB} \tag{8.5}$$

Thus the influence line is the deflected shape of the beam produced by a unit load acting in place of the redundant reaction, divided by a scaling factor equal to the deflection at the point of application of the unit load. This is an example of Müller Breslau's Principle. Alternatively, since at B the value of δ_{XB}/δ_{BB} is unity, the reaction at B due to a unit load at X is equal to the deflection at X due to a unit deflection at B. This is a contragredient transformation which can be extended to problems of multiple redundancies, as will be demonstrated later.

Worked examples

8.5 Reciprocal theorems

A cantilever of length L carries a point load W at the free end. Given that the slope and deflection at the free end are $WL^2/2EI$ and $WL^3/3EI$ respectively,

(a) determine the deflection δ at the free end when the point load is replaced by a clockwise couple M

(b) find the deflection at B in the cantilever of Fig. 8.13(a).

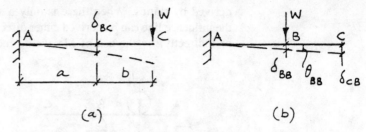

Figure 8.13

Solution (a) The corresponding loads and displacements for the two load-cases are:

Case	Load	Displacement
1	W	$\theta = WL_2/2EI$
2	M	δ

By Betti's theorem,

$$W\delta = M\theta = MWL^2/2EI, \text{ from which } \delta = ML^2/2EI$$

Solution (b) Move the load W to B as in Fig. 8.13(b). Then, according to Maxwell's theorem, since the load is the same in each case, δ_{BC} in Fig. 8.13(a) is equal to δ_{CB} in Fig. 8.13(b). Hence, using the data given,

$$\delta_{BC} = \delta_{CB} = \delta_{BB} + a\theta_{BB} = Wb^2/3EI + aWb^2/EI = Wb^2(2b+3a)/6EI$$

8.6 Maximum deflection

> A beam ABC of uniform section is simply supported at A and B and has a uniform section. The simply supported span AB is 12 m long and the cantilever span BC 3 m.
> (a) Derive the equation of the influence line for the deflection at C.
> (b) Ignoring the self-weight of the beam, determine the maximum upward deflection when a uniformly distributed load of 12 kN/m, 4 m long, crosses the beam. Assume $EI = 60\,000$ kNm^2.

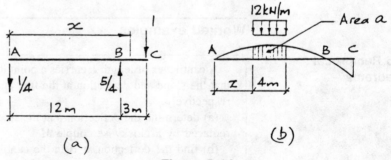

Figure 8.14

Solution – Fig. 8.14 (a) Put the unit load at C and determine the reactions – see Fig. 8.14(a); then apply the differential equation of bending, as follows.

$$Ely'' = x/4 - 5[x - 12]/4$$

where the square brackets indicate a step function. Integrating twice and applying the boundary conditions $y = 0$ when $x = 0$ or 12,

$$y = \{x^3/24 - 5[x - 12]^3/24 - 6x\}/EI$$

This is the equation of the influence line; its shape is shown in Fig. 8.14(b).
Solution (b) Since the load must be on AB to produce an upward deflection the step function vanishes and

$$y = (x^3/24 - 6x)/EI \qquad (1)$$

The area under the load is given by

$$a = 1/EI \times \int_z^{z+4} y\,dx = [(z+4)^4/96 - 3(z+4)^2 - z^4/96 + 3z^2]/EI \quad (2)$$

For maximum deflection $da/dz = 0$.
Hence

$$(z+4)^3/24 - 6(z+4) - z^3/24 + 6z = 0 \qquad (3)$$

i.e. $3z^2 + 12z - 128 = 0$, from which $z = 4.831$ m

Back substituting into equation (2) and putting $EI = 60E3$ gives $a = -1.771E\text{-}3$.
Multiplying by the load intensity,

$$12 \times -1.771E\text{-}3 = -0.0213 \text{ m, i.e. } \delta = 21.3 \text{ mm upwards}$$

Note: As equation (3) can also be obtained from equation (1) by putting $y_z = y_{z+4}$, *the principle of equal ordinates still applies.*

8.7 Propped cantilever

Sketch the influence lines for the force in the prop and the bending moment at the fixed end of a cantilever of length L, propped at the free end.

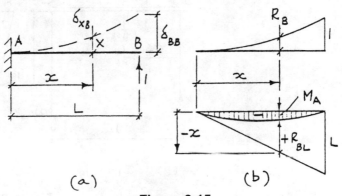

(a)   (b)

Figure 8.15

Solution — Fig. 8.15 (1) Force in prop at B: Apply a unit load in place of the prop and use the differential equation of bending to obtain the deflected shape of the beam.

i.e. $EIy'' = x - L$

Integrating twice and applying the boundary conditions at the fixed end, viz: $x = 0$, $y = 0$, $y' = 0$,

$$\delta_{XB} = y = (x^3/6 - Lx^2/2)/EI$$

which is the deflected shape shown by the dotted line in Fig. 8.15(a). Putting $x = L$

$$\delta_{BB} = -L^3/3EI$$

The equation of the influence line is therefore given by

$$R_B = \delta_{XB}/\delta_{BB} = [3(x/L)^2 - (x/L)^3]/2$$

as shown by the upper diagram in Fig. 8.15(b).

(2) Bending moment at A:

$$M_A = R_B L - x$$

The influence line can therefore be constructed as shown by drawing the straight line $M_A = -x$ and erecting from it the ordinates of the influence line for R_B multiplied by L.

8.8 Portal frame 1

Sketch the influence line for the horizontal reaction at the pinned bases of a portal frame of height H and span L, with constant EI.

Solution — Fig. 8.16 (1) Let the horizontal reaction R_D be the redundant restraint. Then apply unit loads in turn at D and X. The resulting bending moment diagrams m_D and m_X are as shown.

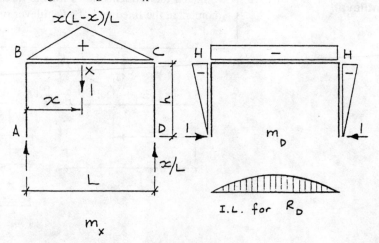

Figure 8.16

(2) From equation (8.5)

$$R_D = \delta_{XD}/\delta_{DD}$$

Using Table 6.1

$$
\begin{aligned}
\delta_{XD} \text{ (upwards)} &= -\Sigma \int m_D \, m_X dz \\
&= 1/2 \times x \times x(L-x)/L \times (-H) \\
&\quad + 1/2 \times (L-x) \times x(L-x)/L \times (-H) \\
&= -H(Lx - x^2)/2
\end{aligned}
$$

$$
\begin{aligned}
\delta_{DD} \text{ (inwards)} &= \Sigma \int m_D^2 \, dz \\
&= 2 \times 1/3 \times H \times (-H)^2 + 1 \times L \times (-H)^2 \\
&= H^2(2H/3 + L)
\end{aligned}
$$

Hence

$$R_D = 3(x/H - x^2/HL)/2(2H/L + 3) \text{ which is maximal when } x = L/2.$$

i.e. $R_D(\text{max}) = 3L/8H(2H/L + 3)$

Since R_D is zero when the load is at B or C, the shape of the influence line is as shown. This must be a segment of a circle as BC is seen to be in pure bending when unit loads are applied at D.

Note: If the unit load method had been used to determine the redundant reaction directly, the form of the equations would have been exactly the same as above.

i.e. $R_D = -\Sigma \int m_D m_X dz \,/\, \Sigma \int m_D^2 dz$

8.9 Portal frame 2

> Derive the influence line for the bending moment at mid-span of the beam BC in the previous problem and use it to find the bending moment due to a point load of 140 kN at the quarter-span position when $L = 20$ m and $H = 5$ m.

Solution – Fig. 8.17 (1) With the unit load at X ($x \le L/2$), the bending moment at mid-span is given by

$$M = x/l \times L/2 - R_D H = x/2 - R_D H$$

The new influence line diagram is therefore a composite of two diagrams, both having maximum values when $x = L/2$, as shown.

(2) Putting $H = 5$, $L = 20$, and using the expression for R_D from the

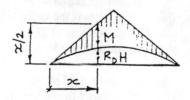

Figure 8.17

previous problem, the ordinate of the new influence line when the load is at the quarter-span position ($x = 5$) is 25/28. Thus, for a 140 kN load

$$M = 140 \times 25/28 = 125 \text{ kNm}$$

Multiple redundancies and model analysis

When there is more than one redundancy, structures with specific load-cases can be analysed directly by the solution of simultaneous equations, as described in Chapter 7. However, with the current easy access to personal computers and the availability of programs for structural analysis it is doubtful whether the analysis by hand of structures with more than two redundancies would be entertained in a structural design office. Even with two redundancies the determination of general expressions enabling the construction of influence lines becomes tedious, and would probably not be undertaken. Structures containing curved members, or members with many section changes or continuously variable sections, are especially difficult to analyse except with the aid of a fairly sophisticated computer program. However, analyses of plane structures, even of the latter type, can often quite easily be carried out with the aid of simple models made from any light elastic sheet material.

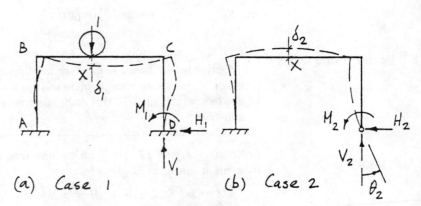

(a) Case 1   (b) Case 2

Figure 8.18

Consider the portal frame in Fig. 8.18(a), for which influence lines for the redundant reactions M_1, H_1, V_1 at the built-in support D are to be determined. Assume that the deflection at point X under the unit rolling load is δ_1. The displacements at D are of course all zero. Now let the support at D of the unloaded frame be replaced by a pin, and the leg at that point be given a rotation θ_2 radians, as in Fig. 8.18(b). The horizontal and vertical displacements at D will be held at zero by the pin; and a new set of reactions M_2, H_2, V_2 will be required to hold the frame in this state. Let the deflection at X now be δ_2. The corresponding loads and displacements for the two cases are set out below.

Case 1	1	M_1	H_1	V_1	δ_1	0	0	0
Case 2	δ_2	θ_2	0	0	0	M_2	H_2	V_2

By Betti's reciprocal theorem

$$1 \times (-\delta_2) + M_1 \times \theta_2 + H_1 \times 0 + V_1 \times 0 = 0 \times \delta_1 + M_2 \times 0 + H_2 \times 0 + V_2 \times 0$$

Hence

$$M_1 = \delta_2/\theta_2 \tag{8.6}$$

Thus the influence line for the redundancy M_1 at D is the deflected shape of the frame when a unit rotation is applied in the direction of M_1 all other displacements at D being held at zero. Similar conclusions can be drawn for the other redundancies by applying unit displacements in the directions of H_1 and V_1 in turn, keeping the other two displacements at zero in each case. This is an extension of Müller Breslau's principle previously described for a single redundancy.

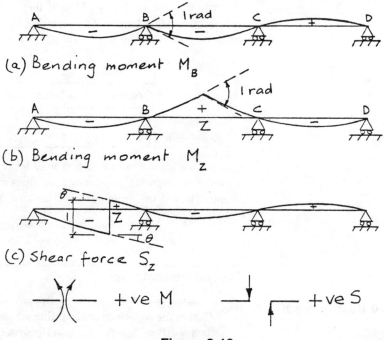

(a) Bending moment M_B

(b) Bending moment M_Z

(c) Shear force S_Z

Figure 8.19

The principle may also be applied to stress resultants, a unit *relative* displacement, made in the positive direction of the stress resultant, producing the influence line, provided that all other possible *relative* displacements at the point are held at zero. Some examples are shown in Fig. 8.19. The meaning of the term *relative* is illustrated in the examples. In case (b) the two parts of the beam at Z are rotated relatively to one another in the positive sense of the bending moment, but there is no longitudinal or transverse separation.

The bodily movement of the point Z itself is not restricted, however, unless, as in case (a), it is restrained by a support. Similarly, in case (c) the two parts of the beam are separated vertically in the positive direction of the shear force. Here there is no relative rotation, so the tangents to the two parts must remain parallel, and again there is no longitudinal separation.

The model is cut out from the sheet material in the shape of the structure, to some convenient scale and attached to a flat board covered with graph paper. Then after a suitable displacement has been made at the point of interest, the deflected shape is transferred to the paper using the deflected model as a guide. Experimental techniques do not need to be very sophisticated to achieve reasonably accurate results.

Second moments of area of the model must be in the same proportions as in the prototype, but can be to any scale. The combination of length scale and second moment of area is chosen so that displacements, sufficient to give deflections that can be accurately measured, can be applied without exerting undue force, while maintaining stability in the model.

The length scale must be taken into account when converting results from the model to the prototype. Thus if the lengths in the prototype are s times the corresponding lengths in the model, the results obtained from the model must be multiplied by the following factors.

	Ordinates of influence line	Areas under influence line
Force	1	s
Moment	s	s^2

Areas under the influence line may be determined with a high degree of accuracy by Simpson's rule which assumes a parabolic curve for the deflected shape. Thus if the ordinates have uniform spacing h and height r,

$$\text{Area} = h/3 \times [r_{\text{first}} + r_{\text{last}} + 2\Sigma r_{\text{odd}} + 4\Sigma r_{\text{even}}] \tag{8.7}$$

Worked examples

8.10 Built-in beam

> Derive the equations of the influence lines for the reactions to a built-in beam AB of length L and uniform section. Use the influence lines to determine the value of the reactions when a uniformly distributed load covers half the beam, from A to mid-span.

Solution − Fig. 8.20 (1) Since the beam is symmetrical, either end may be considered. The reactions at B are V and M, as shown in Fig. 8.20(a).

(2) Remove all the restraints at B. Then by Müller Breslau's principle the influence lines are the deflected shapes of the cantilever when unit displacements are applied in the directions of V and M, as in Fig. 8.20(b) and (c) respectively. In each case the other displacements are held at zero and the beam is subjected

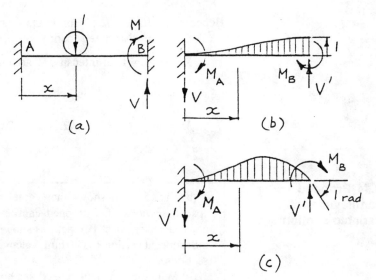

Figure 8.20

to a new, unknown set of reactions V', M_A, M_B. These can be obtained from the slope-deflection equations (4.3) and (4.4), as follows.

$$M_A = 4EI\theta_A/L + 2EI\theta_B/1 + 6EIy/L^2$$
$$M_B = 2EI\theta_A/L + 4EI\theta_B/L + 6EIy/L^2$$

where θ is positive in the direction of the moments and the sign of the last term is changed to make y, and hence the ordinates of the influence line, positive upwards.
Taking moments about one end,

$$V' = (M_A + M_B)/L$$

The deflected shapes may then be derived by means of the differential equation of bending. Since upward deflections are required to be positive

$$EIy'' = M_A - V'x$$

Integrating twice and applying the boundary conditions $y' = y = 0$ when $x = 0$,

$$EIy = M_A x^2/2 - V'x^3/6$$

(3) Influence line for V: apply a unit deflection at B with no rotation, as in Fig. 8.20(b).
Putting $\theta_A = \theta_B = 0$ and $y = 1$ in the slope-deflection equations,

$$M_A = M_B = 6EI/L^2, \quad V' = 12EI/L^2$$

Hence $y = 3x^2/I^2 - 2x^3/L^3$. This is the equation of the influence line for V.
(4) Influence line for M: apply a unit rotation at B with no deflection, as in Fig. 8.20(c).
Putting $\theta_A = 0$, $\theta_B = 1$, $y = 0$ in the slope-deflection equations,

$$M_A = 2EI/L, \quad M_B = 4EI/L, \quad V' = 6EI/L^2$$

Hence $y = x^2/L - x^3/L^2$. This is the equation of the influence line for M.

(5) Effect of distributed load:

Reaction = load intensity × area of influence diagram under the load

i.e. $V = w \int_0^{L/2} (3x^2/L^2 - 2x^3/L^3)dx = 3wL/32$

$M = w \int_0^{L/2} (x^2/L - x^3/L^2)dx = 5wL^2/192$

8.11 Propped cantilever with variable section

(a) Fig. 8.21(a) shows the second moments of area at five equally spaced sections of a propped cantilever. A one tenth scale model is constructed in sheet Perspex, as shown in Fig. 8.21(b). If the depth of the model at section 1 is 20 mm, calculate the required depths at the other sections.

(b) With the end block pinned to a board, an upward force was applied at section 1 producing the deflected shape shown in Fig. 8.21(c). Determine the prop reaction in the prototype when a uniformly distributed load of 40 kN/m covers the whole span.

(c) Determine the ordinates of an influence line for the bending moment at the fixed end of the prototype.

Solution (a) Let I_i and D_i be the second moment of area and depth at section i. Then

$$(D_i/D_1)^3 = I_i/I_1 \quad \text{i.e.} \quad D_i = D_1(I_i/I_1)^{1/3}$$

Hence

$$D_2 = D_1(I_2/I_1)^{1/3} = 20(1.95)^{1/3} = 25 \text{ mm}$$

Similarly

$$D_3, D_4, D_5 = 40, 65, 100 \text{ mm}.$$

Solution (b) Divide all the deflections by 75 to obtain a unit displacement in the direction of the reaction and hence the influence line for the model, as in Fig. 8.21(d). As the reaction is a force, the ordinates for the prototype are the same as for the model, but areas under the line must be multiplied by the scale factor 10. Thus, by Simpson's rule, the area under the influence line for the prototype is

$$A = 0.5/3 \times [1 + 0 + 2 \times 0.136 + 4(0.432 + 0.0453)] \times 10 = 5.302 \text{ m}$$

The area is in m because lengths are in m and the ordinates are influence coefficients whose units are kN/kN. Multiplication by the intensity of the load in kN/m gives the propping force in kN, thus

$$P = 5.302 \times 40 = 212.1 \text{ kN}$$

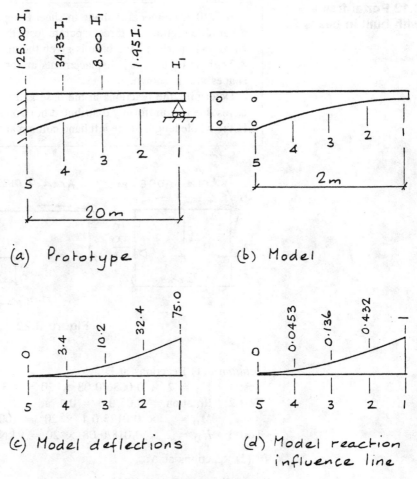

(a) Prototype (b) Model

(c) Model deflections (d) Model reaction influence line

Figure 8.21

Solution (c)   The bending moment in kNm at the fixed end of the prototype, due to a 1 kN rolling load, is given by

$$M = 20P - x$$

where x is the distance of the unit load in m from the fixed end. The force P can be obtained directly from the model influence line. The ordinates at sections 1 to 5 of the bending moment influence line for the prototype are therefore

$$m_i = 0, \ -6.36, \ -7.28, \ -4.09, \ 0$$

In this case the unit of the ordinates is kNm/kN, i.e. m. If the bending moment due to a uniformly distributed load were now required the unit of area under the line would be m^2; and multiplication by the load intensity would give bending moment in kNm.

8.12 Portal frame with built-in bases

A 1/20 scale model is made in sheet Perspex of a portal frame with a variable section. In three separate tests the left hand base is rigidly fixed and the right hand base is given the displacements shown in Fig. 8.22, holding all other displacements at zero. The resulting deflected shapes are as shown.

Determine the reactions at the base and the bending moment at the eaves due to a uniformly distributed horizontal load of 2 kN/m applied to the whole length of the left hand column and acting from left to right.

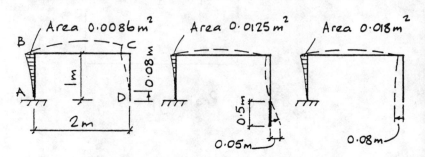

Figure 8.22

Solution   (1) Reactions at D:

Test 1: $V_D = 2 \times 0.0086/0.08 \times 20 = 4.3\,\text{kN}$ upwards

Test 2:   Rotation = $0.05/0.5 = 0.1\,\text{rad}$

$M_D = 2 \times 0.0125/0.1 \times 20^2 = 100\,\text{kNm}$ anticlockwise

Test 3: $H_D = 2 \times 0.018/0.08 \times 20 = 9\,\text{kN}$ from right to left

(2) Reactions at A:

$V_A = -V_D = 4.3\,\text{kN}$ downwards

$H_A = \text{load} - H_D = 2 \times 20 - 9 = 31\,\text{kN}$ from right to left

$M_A = -M_D - 40V_D + 40 \times 10 = -128\,\text{kNm}$ anticlockwise

(3) Bending moment at eaves:

$M_C = -20H_D + M_D = -80\,\text{kNm}$

$M_B = M_C + 40V_D = 92\,\text{kNm}$

Problems

1   A symmetrical 3-pinned parabolic arch has span $2L$ and height h.

(a) Draw influence lines for the horizontal and vertical reactions at a support.

(b) Derive an influence line for the bending moment at the quarter-span positions and prove that when a long uniformly distributed load crosses the span the maximum positive and negative bending moments there have the same numerical value.

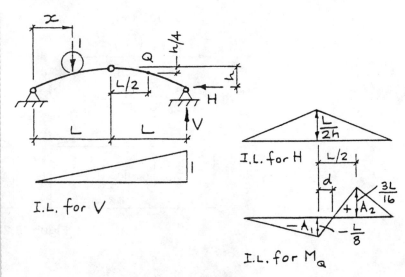

Figure 8.23

Solution − Fig. 8.23

  (a) *V*: Since H has no moment about A the influence line is as shown.

      H: Load at supports: $H = 0$

           Load at crown: $V = 1/2,\ H = VL/h = L/2h$

  (b) *M*: Load at supports: $M = 0$

           At Q: $x = 3L/2$, height $= 3h/4$ (property of parabola)

           Load left of Q: $M = VL/2 - 3hH/4$

           Load at Q: $V = 3/4,\ H = L/4h,\ M = 3L/16$

           Load at crown: $V = 1/2,\ H = L/h,\ M = -L/8$

      From influence line for M:

$$d = L/5,\ L+d = 6L/5,$$
$$L-d = 4L/5$$
$$A_1 = 6L/5 \times L/16 = 3L^2/40$$
$$A_2 = 4L/5 \times 3L/32 = 3L^2/40 = A_1$$
$$\text{(Q.E.D.)}$$

2   A simply supported beam with a span of 60 m is traversed by a train of four loads, viz: 6, 14, 10, 10 kN, spaced 10 m apart. The train may travel in either direction, with the 10 kN load leading.

    Use influence lines to determine the maximum values of the shear force 12 m from the left hand support and the bending moment 20 m from the same support.

Solution − Fig. 8.24

    Draw the influence lines and, as one of the loads must be at the point of interest, calculate the ordinates at 10 m spacing, as shown, by simple

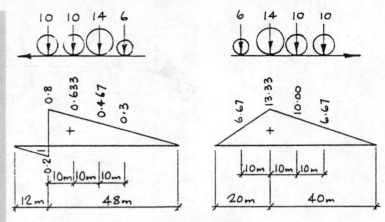

Figure 8.24

proportion. Determine, by inspection or trial and error, the position and direction of the train that give maximum values.

$S_{max} = 10 \times 0.8 + 10 \times 0.633 + 14 \times 0.467 + 6 \times 0.3 = 22.67 \text{ kN}$

$M_{max} = 10 \times 6.67 + 10 \times 10.0 + 14 \times 13.33 + 6 \times 6.67 = 393.3 \text{ kNm}$

3 A symmetrical bridge is in the form of two cantilever beams 15 m long, connected at their free ends by a central simply supported span of 20 m.

Draw the following influence lines: for a cantilever the shear force and bending moment at the built-in end; for the simply supported span the shear force at a support and the bending moment at mid-span. Determine the maximum bending moment at a built-in end when a uniformly distributed load of 10 kN/m, 14 m long, crosses the bridge.

Solution — *Fig. 8.25*

(1) Define the influence lines by determining the stress resultants s_A, m_A, s_B, m_C, with the unit load at A, B, D, E; and also at C for m_C.

(2) For M_A(max) determine x and y to give equal ordinates m on either side of B:

i.e. solving $m = -15 + x = -15(20 - y)/20$ and $x + y = 14$ gives $x = 6$, $y = 8$, $m = -9$

Hence M_A(max) $= -10(6 + 8)(9 + 15)/2 = -1680 \text{ kNm}$

4 A truss consists of nine equilateral triangles of 3 m side arranged as shown in Fig. 8.26. Assuming that loads travel along the top boom, draw influence lines for the member forces X, Y, and Z. Use the influence line to determine the maximum compressive value of Y when a uniformly distributed load of 50 kN/m and 7.5 m long crosses the truss.

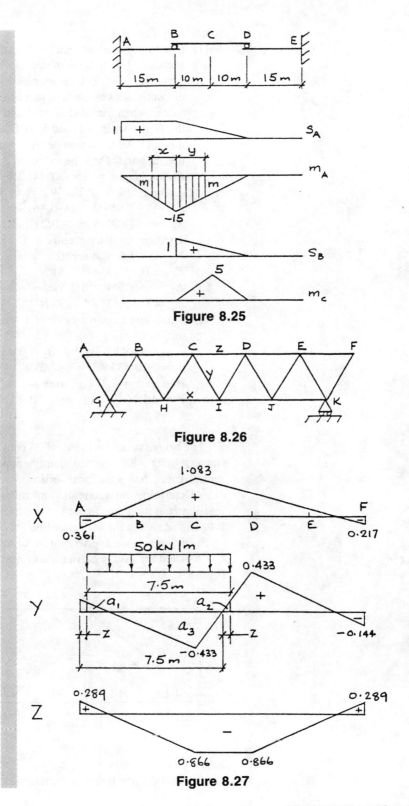

Figure 8.25

Figure 8.26

Figure 8.27

Solution — Fig. 8.27

(1) Cut the frame through the members and use the method of sections. For X take moments about C; for Y resolve forces vertically; for Z take moments about I. There are possible discontinuities as the load passes C and D, so determine member forces when the load is at C and D. Use the left hand part of the frame when the load is at D, and the right hand part when it is at C. All three member forces are zero when the load is above the supports. Find the values at A and F by simple proportion.

Frame height $= 3\cos 30° = 2.598$ m

$X_C = 1.5/4 \times 7.5/2.598 = 1.083$ kN

$X_A = -1.083/3 = -0.361$ kN

$X_F = -1.083/5 = -0.217$ kN

There is no discontinuity at D

$Y_C = -1.5/4 \times \sec 30° = -0.433$ kN $\qquad Y_D = -Y_C$

$Y_A = 0.433/3 = 0.144$ kN $\qquad\qquad\quad Y_F = -Y_A$

$Z_C = -1.5/4 \times 6/2.598 = -0.866$ kN $\quad Z_D = Z_C$

$Z_A = 0.866/3 = 0.289$ kN $\qquad\qquad\quad Z_F = Z_A$

(2) Maximum negative value of Y occurs when area $(a_1+a_2+a_3)$ is minimal.

$a_1 = 0.144(1.5-z)^2/3$, $a_2 = 0.433z^2/3$

$a_3 = -0.433 \times 3 = -1.299$

Hence: $z = 0.374$ m, area $= -1.218$ m

$Y = 50 \times (-1.218) = -60.9$ kN

5 The bridge structure in Fig. 8.28 consists of a central rigidly jointed frame ABCD with attached simply supported spans EB and CF. The central frame has a uniform section.

(a) Determine the maximum and minimum values of the horizontal reaction H at the pinned bases of the frame as a unit load crosses from B to C, and sketch the influence line for the whole length of the bridge.

(b) Obtain a close approximation to the bending moment at B when $L = 5$ m and the bridge carries a uniformly distributed load of 20 kN/m from E to C.

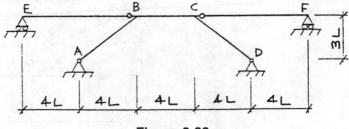

Figure 8.28

Solution — Fig. 8.29

(a) On BC, maximum and minimum values of H occur when the

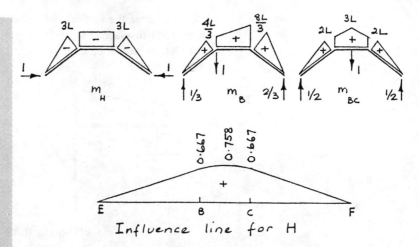

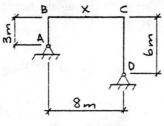

Influence line for H

Figure 8.29

rolling load is at mid-span or the end of BC. Use the standard unit load method to find the redundant reaction H at A. Draw bending moment diagrams m_H (unit horizontal load at A), m_B (unit vertical load at B), m_{BC} (unit vertical load on BC), as shown. Use Table 6.1 to evaluate the integrals.

$\Sigma\int m_H m_B dz = -44L^3$, $\Sigma\int m_H^2 dz = 66L^3$, $H = 44/66 = 0.667$ with load at B or C

$\Sigma\int m_H m_{BC} dz = -50L^3$, $H = 50/66 = 0.758$ with load at mid-span of BC

The influence line is as shown, with straight lines for the statically determinate end spans.

(b) H = load intensity × area under influence line from C to E
$\approx 20[8L \times 0.667 + 2/3 \times 4L(0.758-0.667)] = 588$ kN
for $L = 5$ m

Vertical reaction at A = 400(1/2 + 2/3) = 467 kN

Bending moment at B = 467 × 20 − 558 × 15 = 970 kNm

6 The pinned-base portal frame ABCD in Fig. 8.30 has constant EI. Point X is at mid-span on BC.

Figure 8.30

(a) Determine the vertical deflection at X when the base at D is given a horizontal displacement of 200 mm.

(b) Determine the bending moment at X when a point load of 30 kN is placed there.

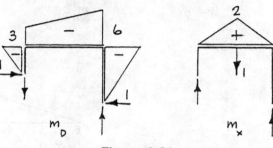

Figure 8.31

Solution — Fig. 8.31

By Müller-Breslau the vertical deflection at X due to a unit horizontal displacement at D is equal to the horizontal reaction at D due to a unit vertical load at X. Remove the horizontal reaction H_D at D. Then apply unit loads at D and X to give bending moments m_D and m_X, as shown. Use Table 6.1.

(a) $\sum \int m_D m_X \, dz = -36$, $\sum \int m_D^2 = 249$, $H_D = 36/249 = 0.1446$

for unit load at X

Hence $\delta_X = 200 \times 0.1446 = 28.9$ mm

(b) $H_D = 30 \times 0.1446 = 4.338$ kN for 30 kN load at X

Hence $V_D = 15 + 4.338 \times 3/8 = 16.63$ kN,

$M_X = 16.63 \times 4 - 4.338 \times 6 = 40.5$ kNm

7 A 1/20 scale model of a continuous beam ABCD is cut and pinned at B. In a test the two parts of the model at this point were rotated relatively by means of rigid extensions. The extensions were then set to include an angle of 155° and the model allowed to deflect freely, as shown in Fig. 8.32. The ordinates in mm of the deflected shape, measured at 200 mm intervals, starting from A, were: 0, 21.8, 35.3, 29.2, 0, 24.9, 28.8, 14.4, 0, 4.4, 3.1, 0, in the directions indicated in the figure.

Determine the maximum bending moment at B and at mid-span in AB of the prototype beam when a long uniformly distributed load of 2.4 kN/m crosses the beam.

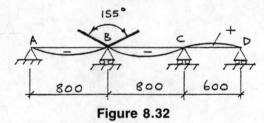

Figure 8.32

Solution

For M_B(max): AB and BC loaded, CD unloaded; for M_{AB}(max): AB only loaded

Areas, by Simpson's rule, are:

AB: $-200/3 \times [2 \times 35.3 + 4(21.8 + 29.2)] \times 1E\text{-}6 = -0.0183 \text{ m}^2$

BC: $-200/3 \times [2 \times 28.8 + 4(24.9 + 14.4)] \times 1E\text{-}6 = -0.0143 \text{ m}^2$

Total area $= -0.0326 \text{ m}^2$

M_B(max) $= -2.4 \times 0.0326 \times 20^2 \times 180/\pi(180 - 155)$

$\qquad = -71.7 \text{ kNm}$

M_B (AB only loaded) $= -71.7 \times 0.0183/0.0326 = -40.3 \text{ kNm}$

M_{AB}(max) $= 2.4 \times 16^2/8 - 40.3/2 = 56.7 \text{ kNm}$

8 A simply supported 3-span continuous beam ABCD has spans AB = CD = 15 m, BC = 30 m. The section is uniform throughout. The loading is uniformly distributed and can vary on any span between 12 kN/m and 4 kN/m. Determine, from the following tests on a model, the maximum reaction and bending moment at B, and the maximum bending moment at mid-span in BC.

The model is 1/30 scale and is simply supported at A, C, and D. An upward deflection at B produced the following deflections in mm at 250 mm intervals.

$\quad$ 0(A), 61.7, 100(B), 98.4, 68.8, 29.7, 0(C), -7.0, 0(D)

Solution − Fig. 8.33

$\quad$ (1) By Simpson's Rule, model areas in m are $A_1 = 0.0289$, $A_2 = 0.0625$, $A_3 = -0.00233$

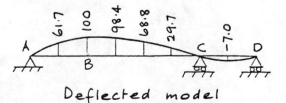

Deflected model

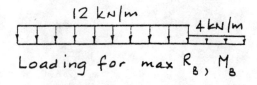

Loading for max R_B, M_B

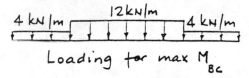

Loading for max M_{BC}

Figure 8.33

(2) For prototype multiply areas by $30 \times 1/0.1$, i.e.: $A_1 = 8.67$, $A_2 = 18.75$, $A_3 = 0.70$ m

(3) Using the loading for maximum R_B and M_B (see Fig. 8.33),
$$R_B = 12(A_1 + A_2) + 4A_3 = 326 \text{ kN}$$

(4) Reversing the influence line for R_C, since the beam is symmetrical, $R_C = 12(A_3 + A_2) + 4A_1 = 251$ kN. By statics $R_A = 37.75$ kN, $M_B = -784$ kNm

(5) Using the loading for maximum M_{BC} (see Fig. 8.33),
$$R_B = R_C = 4A_1 + 12A_2 + 4A_3 = 257 \text{ kN}$$
By statics $R_A = -17$ kN, $M_{BC} = 645$ kNm

9

Plastic method

Introduction

Plastic analysis is a hand method of structural analysis which determines the forces and moments in certain types of structures where members are predominantly in bending at collapse. The method is generally limited to structures in one plane and assumes that axial and shear forces have little effect on the solution of a structure which is essentially in bending.

The method must only be applied to structures where the material becomes plastic at the yield stress and is capable of accommodating large plastic deformations. It must therefore not be applied to brittle materials such as cast iron. However, it can be applied with restraints to reinforced concrete because the steel reinforcement behaves plastically at collapse.

The plastic method can be seen as a more rational method for design because all structures can be given the same safety factor against collapse. In contrast for elastic methods the safety factor varies. However, calculations for instability and elastic deflections require careful consideration when using the plastic method, but nevertheless it is very popular for the design of some structures, e.g. portal frames.

Intrinsically the plastic method of analysis is simpler than the elastic method because there is no need to satisfy elastic strain compatibility conditions and also there are no simultaneous equations to solve as in the slope deflection method.

The method demonstrated in this chapter is based on the principle of virtual work. This states that if a structure, which is in equilibrium, is given a set of small displacements then the work done by the external loads on the external displacements is equal to the work done by the internal forces on the internal displacements. More concisely, external work equals internal work. The displacements need not be real; they can be arbitrary, which explains the use of the word 'virtual'. However, the external and internal geometry must be compatible as illustrated in the following problems.

It is tacitly assumed for examples in this book that collapse is due to the formation of plastic hinges at certain locations and that other possible causes of failure, e.g. local or general instability, axial or shear forces, do not occur. It is also important to understand that at collapse:

(1) the structure is in equilibrium i.e. the forces and moments, externally and internally, balance;

(2) no bending moment exceeds the plastic moment of resistance of a member;

(3) there are sufficient hinges to form a collapse mechanism.

These three conditions lead to three theorems for plastic analysis.

Lower bound theorem: if only conditions (1) and (2) are satisfied then the solution is less than or equal to the collapse load.

Upper bound theorem: if only conditions (1) and (3) are satisfied then the solution is greater than or equal to the collapse load.

Uniqueness theorem: if conditions (1), (2) and (3) are satisfied then the solution is equal to the collapse load.

Settlement of the supports has no effect on the solution at collapse because the only effect is to change the amount of rotation required. This is in contrast to elastic methods of analysis where settlement calculations must be included.

Plastic hinges form in a member at the maximum bending moment. However, at the intersection of two members, where the bending moment is the same, the hinge forms in the weaker member. Generally the locations of hinges are at restrained ends, intersection of members and at point loads. The hinges may not form simultaneously as loading increases but this is not important for calculating the final collapse load. Generally the number of plastic hinges is given by

$$n = r + 1$$

where r is the number of redundancies. However, there are exceptions, e.g. partial collapse of a beam in a structure.

In some problems the term shape factor, s, which is the ratio of elastic to plastic section modulus, is used. The shape factor varies dependent on the cross section, e.g. for a rectangular section $s = 1.5$, and for steel 'I' sections in common use in design $s = 1.15$. In other situations the term load factor, λ, which is the ratio of the load at plastic collapse to service load, is used. The load factor generally varies between 1.5 and 2.5.

No sign convention is required for the virtual work method, but the sign convention for plotting the bending moment diagrams is based on the right hand screw rule as described in Volume 1 of *Solving Problems in Structures*.

Worked examples

9.1 Elastic and plastic moment of resistance

Compare the elastic and plastic moments of resistance of a simply supported beam carrying a central point load as shown in Fig. 9.1. The beam cross section is rectangular.

Solution If the load W is gradually increased from zero to collapse of the beam then the appearance of the beam and the distribution of bending moments is as shown in Fig. 9.1(a). At mid-span, where the bending moment is a maximum, the vertical distribution of bending stress varies from elastic to elastic-plastic to plastic, as shown in Fig. 9.1(d). The stress in any fibre of

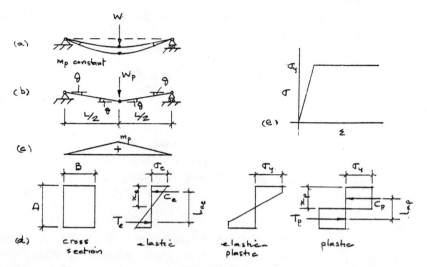

Figure 9.1

the beam changes from elastic to plastic when the stress reaches yield (σ_y) as shown in Fig. 9.1(e).

In the elastic stage of behaviour the tensile and compressive forces are in equilibrium, and for a rectangular section (see Fig. 9.1(d))

$$C_e = T_e$$

and $Bx_e\sigma_e/2 = B(D-x_e)\sigma_e/2$; hence $x_e = D/2$

In general in the elastic stage of behaviour the neutral axis of bending is at the centroid of the section.

The elastic moment of resistance for a rectangular section is

$$m_e = C_e l_{ae}$$
$$= B(D/2)(\sigma_e/2)(2D/3)$$
$$= (BD^2/6)\sigma_e$$

The quantity $(BD^2/6) = z_e$ is the elastic section modulus for a rectangular cross section. The alternative and more general method of obtaining z_e is from the relationship $z_e = I/y$, where I is the second moment of area about the neutral axis and y is the distance to the extreme fibres (see Chapter 6 of *Solving Problems in Structures*, Volume 1).

In the plastic stage the tensile and compressive forces are also in equilibrium (see Fig. 9.1(d)), i.e.

$$C_p = T_p$$

and $Bx_p\sigma_y = B(D-x_p)\sigma_y$; hence $x_p = D/2$

In general the neutral axis of bending is at a position where the compressive area is equal to the tensile area, because in most situations the compressive yield stress is equal to the tensile yield stress. It should be emphasised that this is not the centroidal axis except for sections symmetrical about the xx axis, e.g. rectangular and circular sections.

The plastic moment of resistance is

$$m_p = C_p I_{ap}$$
$$= B(D/2)\sigma_y(D/2)$$
$$= (BD^2/4)\sigma_y$$

The quantity $(BD^2/4) = z_p$ is the plastic section modulus for a rectangular cross section. In general where the compressive and tensile yield stresses are equal the plastic section modulus is the first moment of area of the section about the axis of bending.

The shape factor $s = z_p/z_e$ and for a rectangular section

$$s = (BD^2/4)/(BD^2/6) = 1.5$$

If the central point load W on a simply supported beam is increased the beam deflects in the elastic stage until a plastic hinge forms at the section of maximum bending moment, i.e. mid-span (see Fig. 9.1(c)). Equating the external bending moment to the internal plastic moment of resistance

$$W_p L/4 = m_p; \text{ hence } W_p = 4\dot{m}_p/L$$

An alternative method of determining the collapse load, which is of more general use, is to use the principle of virtual work. When the plastic hinge forms the deflection increases with no increase in load and the angle of the beam changes (θ) as shown in Fig. 9.1(a). This change in angle in the plastic stage is idealised for calculation purposes as shown in Fig. 9.1(b). Real hinges are shown as open circles and plastic hinges as black circles. Work is done in rotating the plastic hinge at mid-span but no work is done at the real hinges at the supports because they are assumed to be frictionless. Using the principle of virtual work

external work of the load = internal work at the plastic hinge

Therefore

$$W_p(L/2)\theta = m_p(\theta+\theta); \text{ hence } W_p = 4m_p/L$$

This is the same as the answer obtained previously.

9.2 Elastic and plastic moment of resistance of an 'I' section

> Compare the elastic and plastic moment of resistance of a simply supported beam carrying a point load at quarter span. The beam has an 'I' cross section as shown in Fig. 9.2(a).

Solution The neutral axis of bending is at half the depth for the elastic and plastic conditions because the section is symmetrical about the xx axis. The elastic section modulus (second moment of area about the neutral axis divided by the distance to the extreme fibres) is given by

$$z_e = I_e/(D/2) = [BD^3/12 - (B-t)(D-2T)^3/12]/(D/2)$$
$$= [100 \times 200^3/12 - (100-5)(200-2 \times 10)^3/12]/(200/2)$$
$$= 204.97E3 \text{ mm}^3$$

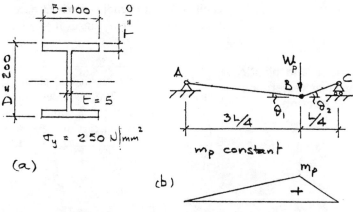

Figure 9.2

The plastic section modulus (first moment of area about the neutral axis) is given by

$$z_p = 2BT(D-T)/2 + 2t(D/2-T)^2/2$$
$$= 2 \times 100 \times 10 \times (200-10)/2 + 2 \times 5 \times (200/2-10)^2/2$$
$$= 230.5\text{E}3 \text{ mm}^3$$

The shape factor is

$$z_p/z_e = 230.5\text{E}3/204.97\text{E}3 = 1.125$$

If the load at quarter span on the simply supported beam is increased until a plastic hinge forms at the point load, then the idealised deflection of the beam in the plastic stage of behaviour is shown in Fig. 9.2(b). Notice that the angles of rotation (θ_1 and θ_2) are given adjacent to the plastic hinge. Using the principle of virtual work

external work = internal work

Therefore

$$W(3L/4)\theta_1 = m_p(\theta_1+\theta_2) \tag{1}$$

From the geometry of the deformed beam in Fig. 9.2(b)

$$(3L/4)\theta_1 = (L/4)\theta_2$$

Rearranging

$$3\theta_1 = \theta_2 \tag{2}$$

Combining (1) and (2) and rearranging

$$W_p = 16m_p/(3L)$$
$$= 16 \times 230.5\text{E}3 \times 250 \times 1\text{E-}3/(3 \times 4\text{E}3)$$
$$= 76.83 \text{ kN}$$

The bending moment diagram is shown in Fig. 9.2(b). The bending moments and the mechanism meet the three requirements for plastic collapse, i.e.

equilibrium, no bending moment exceeds m_p, and a feasible mechanism is used. The number of hinges $n = r+1 = 0+1 = 1$, which is correct.

9.3 Elastic and plastic moment of resistance of a 'T' section

Determine the elastic and plastic section moduli of the 'T' section shown in Fig. 9.3(a). Also determine the load at plastic collapse if the 'T' section is used as a simply supported beam carrying a uniformly distributed load on part of the span as shown in Fig. 9.3(b).

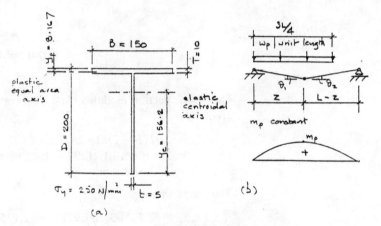

Figure 9.3

Solution The neutral axis of bending is not in the same position for the elastic and plastic conditions for a 'T' section. The area of the 'T' section (see Fig. 9.3(a)) is

$$A = BT + t(D-T)$$
$$= 150 \times 10 + 5 \times (200-10) = 2.45E3 \text{ mm}^2$$

Taking moments of areas about an axis at the bottom of the section (see Fig. 9.3(a)) to determine the position of the elastic centroidal axis gives

$$BT(D-T/2) + t(D-T)^2/2 = Ay_e$$

Therefore

$$150 \times 10 \times (200-10/2) + 5 \times (200-10)^2/2 = 2.45E3y_e$$

Hence

$$y_e = 156.2 \text{ mm from the bottom of the section}$$

Elastic second moment of area about the centroidal axis ($I = BD^3/3$ for a rectangle about one edge, (see Chapter 6 of *Solving Problems in Structures*, Volume 1)

$$I_e = B(D-y_e)^3/3 - (B-t)(D-T-y_e)^3/3 + ty_e^3/3$$

$$= 150(200-156.2)^3/3-(150-5)(200-10-156.2)^3/3+5\times$$
$$156.2^3/3$$
$$= 8.687\text{E}6 \text{ mm}^4$$

The elastic section modulus for the bottom of the section is

$$z_e = I_e/y_e$$
$$= 8.687\text{E}6/156.2 = 55.61\text{E}3 \text{ mm}^3$$

The elastic section modulus for top of the section is

$$z_e = I_e/(D-y_e)$$
$$= 8.687\text{E}6/(200-156.2) = 198.3\text{E}3 \text{ mm}^3$$

The position of the plastic equal area axis of bending is obtained from

area in compression = area in tension

Therefore

$$By_p = t(D-T)+B(T-y_p)$$

and $150y_p = 5(200-10)+150(10-y_p)$

Hence

$$y_p = 8.167 \text{ mm from the top of the section}$$

The plastic section modulus (first moment of area about the plastic equal area axis of bending) is

$$z_p = By_p^2/2+(B-t)(T-y_p)^2/2+t(D-y_p)^2/2$$
$$= 150\times8.167^2/2+(150-5)(10-8.167)^2/2+5(200-8.167)^2/2$$
$$= 97.245\text{E}3 \text{ mm}^3$$

The plastic moment of resistance is

$$m_p = z_p\sigma_y = 97.245\text{E}3\times250 = 24.31\text{E}6 \text{ Nmm}$$

To determine the collapse load (see Fig. 9.3(b)) use the principle that

external work = internal work

Therefore

$$w_pz^2\theta_1/2+w_p(L-z)^2\theta_2/2-w_p(L/4)^2\theta_2/2 = m_p(\theta_1+\theta_2) \qquad (1)$$

From the geometry of the deflected beam (see Fig. 9.3(b))

$$z\theta_1 = (L-z)\theta_2 \qquad (2)$$

Combining (1) and (2) and rearranging

$$w_p/(2m_p) = 1/(15Lz/16-z^2) \qquad (3)$$

The left hand side of equation (3) is a minimum when $dw_p/dz = 0$, i.e.

$$-(15L/16-2z) = 0; \text{ hence } z = 15L/32 \qquad (4)$$

Substituting (4) in (3) and rearranging

$$w_p = 2048m_p/(225L^2)$$
$$= 2048\times24.31\text{E}6/(225\times2\text{E}3^2) = 55.32 \text{ N/mm}$$

The bending moment diagram is shown in Fig. 9.3(b).

9.4 Mast with a varying solid circular section

Determine the plastic collapse load in bending for a mast loaded with a horizontal load W at the top (see Fig. 9.4). The mast is a solid circular section varying in diameter from D_1 at the top to $2D_1$ at the bottom.

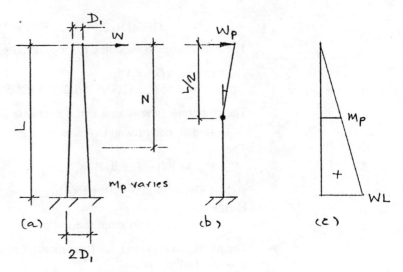

Figure 9.4

Solution The plastic section modulus for a solid circular section (first moment of semi-circular areas about the equal area axis of bending) is

$$z_p = 2(\pi D^2/8)(4/3\pi)(D/2) = D^3/6$$

From the virtual work equation at a distance z from the top of the mast (see Fig. 9.4(b))

external work = internal work

Therefore

$$W_p z\theta = m_p \theta \tag{1}$$

and from the geometry of the taper on the mast

$$m_p = D_1^3(1+z/L)^3 \sigma_y \tag{2}$$

combining (1) and (2)

$$W_p = D_1^3(1+z/L)^3 \sigma_y /(6z) \tag{3}$$

The left hand side of (3) will be a minimum when $dW_p/dz = 0$, i.e.

$$1+z/L = 3z/L; \text{ hence } z/L = 0.5 \tag{4}$$

Combining (3) and (4)

$$W_p = 1.125 D_1^3 \sigma_y /L$$

The bending moment diagram is shown in Fig. 9.4(c).

9.5 Hyperstatic beams

Determine the plastic collapse loads for the three hyperstatic beams shown in Fig. 9.5.

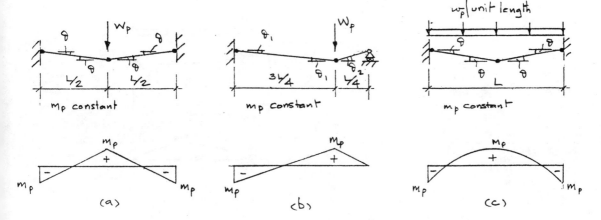

Figure 9.5

Solution (a) — Fig. 9.5(a)

external work = internal work

Therefore

$$W_p L\theta/2 = 4m_p\theta; \text{ hence } W_p = 8m_p/L$$

According to the formula $(n = r + 1)$, the number of plastic hinges should be $n = 3 + 1 = 4$. Only three are necessary and this example is an exception in the use of the formula.

Solution (b) — Fig. 9.5(b)

external work = internal work

Therefore

$$3W_p L\theta_1/4 = 2m_p\theta_1 + m_p\theta_2 \tag{1}$$

According to the formula $(n = r + 1)$, the number of plastic hinges should be $n = 1 + 1 = 2$. There are two which is correct.

From the geometry of the collapse mechanism

$$\theta_2 = 3\theta_1 \tag{2}$$

Combining (1) and (2)

$$W_p = 20m_p/(3L)$$

Solution (c) — Fig. 9.5(c)

external work = internal work

Therefore

$$w_p L^2\theta/4 = 4m_p\theta; \text{ hence } w_p L = 16m_p/L$$

The bending moment diagrams are shown below the beams.

9.6 Propped cantilever

Determine the load at plastic collapse for the propped cantilever carrying a uniformly distributed load shown in Fig. 9.6(a).

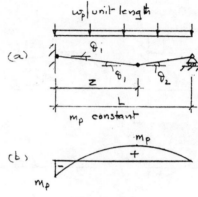

Figure 9.6

Solution To determine the collapse load (see Fig. 9.6(a))

external work = internal work

Therefore

$$w_p z^2 \theta_1/2 + w_p(L-z)^2 \theta_2/2 = 2m_p\theta_1 + m_p\theta_2 \qquad (1)$$

From the geometry of the deflected beam

$$z\theta_1 = (L-z)\theta_2 \qquad (2)$$

Combining (1) and (2)

$$w_p L/(2m_p) = (2L-z)/(Lz-z^2) \qquad (3)$$

The left hand side of the equation will be a minimum when $dw_p/dz = 0$, i.e.

$$(2L-z)/(Lz-z^2) = -1/(L-2z); \text{ hence } z/L = 2-/2 \qquad (4)$$

Substituting (4) in (3) and rearranging

$$w_p = 2m_p/[(3-2\sqrt{2})L^2] $$

The bending moment diagram is shown in Fig. 9.6(b).

9.7 Two span continuous beam

Determine the loads at plastic collapse for the continuous two span beam shown in Fig. 9.7(a). If $W_p = w_p L_2$ determine the condition for the simultaneous collapse of the two spans.

Solution A collapse mechanism may form independently in either span. For span AB equating external and internal work

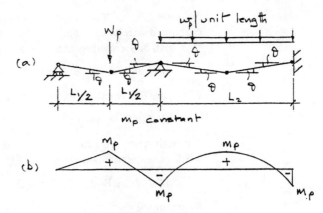

Figure 9.7

$$W_p L_1 \theta/2 = 3m_p\theta; \text{ hence } W_p = 6m_p/L_1 \tag{1}$$

For span BC equating external and internal work

$$w_p L_2^2 \theta/4 = 4m_p\theta; \text{ hence } w_p = 16m_p/L_2^2 \tag{2}$$

If $W_p = w_p L_2$ and the spans collapse simultaneously then equating (1) and (2)

$$6m_p/L_1 = 16m_p/L_2; \text{ hence } L_2/L_1 = 8/3$$

The bending moment diagram is shown in Fig. 9.7(b).

9.8 Cancellation of plastic hinges

Determine the loads at collapse for the continuous two span beam shown in Fig. 9.8(a).

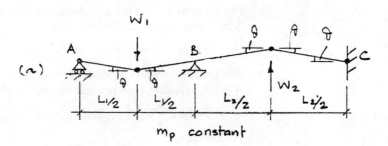

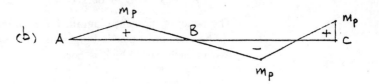

Figure 9.8

Solution A collapse mechanism may form in either span, or there may be a combined mechanism and a cancellation of hinges at support B.

For span AB equating external and internal work

$$W_1 L_1 \theta/2 = 3m_p\theta; \text{ hence } W_1 = 6m_p/L_1 \tag{1}$$

For span BC equating external and internal work

$$W_2 L_2 \theta/2 = 4m_p\theta; \text{ hence } W_2 = 8m_p/L_2 \tag{2}$$

For both spans with cancellation of hinges at B equating external and internal work (see mechanism shown in Fig. 9.8(a))

$$W_1 L_1 \theta/2 + W_2 L_2 \theta/2 = 5m_p\theta$$

Rearranging

$$W_1 + W_2 L_2/L_1 = 10m_p/L_1 \tag{3}$$

The bending moment diagram is shown in Fig. 9.8(b).

9.9 Rectangular sway frame

Determine the loads at collapse for the sway frame shown in Fig. 9.9(a). If $W_1 = W_2 = W$ and $L = 2H$ determine the critical mechanism.

Solution A collapse mechanism may form in beam BC, or a sway mechanism in member AB, or there may be a combined mechanism and a cancellation of hinges at B.

For the beam mechanism in member BC (see mechanism shown in Fig. 9.9(a)), equating external and internal work

$$W_1 L\theta/2 = 3m_p\theta; \text{ hence } W_1 = 6m_p/L \tag{1}$$

For the sway mechanism in member AB (see mechanism shown in Fig. 9.9(b)), equating external and internal work

$$W_2 H\theta = 2m_p\theta; \text{ hence } W_2 = 2m_p/H \tag{2}$$

For both spans with cancellation of hinges at B (see mechanism shown in Fig. 9.9(c)), equating external and internal work

$$W_1 L\theta/2 + W_2 H\theta = 2m_p\theta + m_p\theta$$

Rearranging

$$W_1 + 2W_2 H/L = 6m_p/L \tag{3}$$

The bending moment diagram for each case is shown in Fig. 9.9.

The number of plastic hinges should be $n = r + 1 = 1 + 1 = 2$. There are two in each case which is correct.

If $W_1 = W_2 = W$ and $L = 2H$ then from (1) $W = 6m_p/L$, from (2) $W = 4m_p/L$ and from (3) $W = 3m_p/L$. The value from (3) is the least value and therefore the mechanism shown in Fig. 9.9(c) is critical.

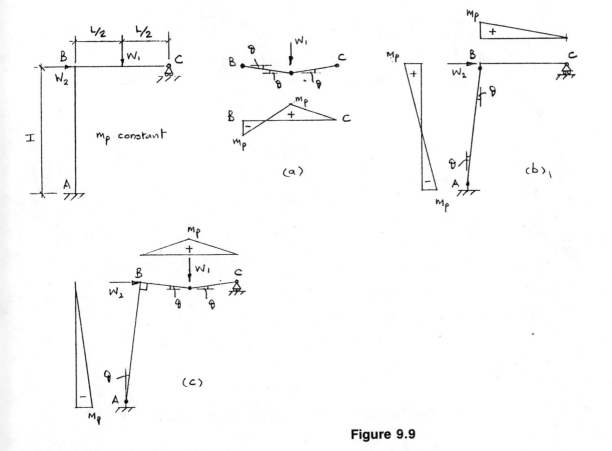

Figure 9.9

9.10 Inclined member

Determine the load at collapse for the sway frame shown in Fig. 9.10(a).

Solution There are no beam mechanisms in this problem because the point loads are applied at the joint. However, there is a sway mechanism.

The work done by a point load at the end of an inclined member is obtained as follows. If member AB rotates through an angle θ then point B moves to B′ as shown in Fig. 9.10(b) and shown enlarged in Fig. 9.10(c).
The work done by the vertical load V is

$$V(B0) = V(BB'\cos\alpha) = VR\theta\cos\alpha = V(L_1/\cos\alpha)\theta\cos\alpha = VL_1\theta$$

The work done by the horizontal load P is

$$P(B'0) = P(BB'\sin\alpha) = PR\theta\sin\alpha = P(H/\sin\alpha)\theta\sin\alpha = PH\theta$$

These results are important and are used frequently in the following examples.
For the sway mechanism (see Fig. 9.10(b)), equating external and internal work and incorporating the previous relationships

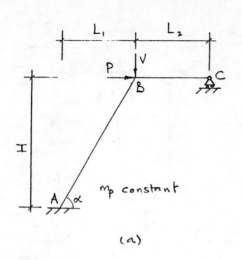

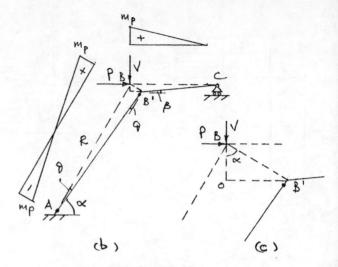

(a)

(b)

(c)

Figure 9.10

$$VL_1\theta + PH\theta = 2m_p\theta + m_p\beta \qquad (1)$$

From the geometry

$$L_2\beta = L_1\theta; \text{ hence } \beta = \theta(L_1/L_2) \qquad (2)$$

Combining (1) and (2)

$$V + PH/L_1 = (m_p/L_1)(2 + L_1/L_2)$$

The bending moment diagram is shown in Fig. 9.10(b).

9.11 Non-rectangular sway frame

Determine the loads at collapse for the sway frame shown in Fig. 9.11(a). If service loads $W_1 = 20$ kN and $W_2 = 10$ kN, $L_1 = L_2 = H = 4$ m, and $m_p = 60$ kNm, determine the lowest load factor for collapse.

Solution A collapse mechanism may form in beam BC, or a sway mechanism in member AB, or there may be a combined mechanism and a cancellation of hinges at B.

For the beam mechanism in member BC (see mechanism shown in Fig. 9.11(b)), equating external and internal work

$$W_1 L\theta/2 = 3m_p\theta; \text{ hence } W_1 = 6m_p/L_1 \qquad (1)$$

For the sway mechanism in member AB (see mechanism shown in Fig. 9.11(c)), equating external and internal work

$$W_2 H\theta + W_1 L_2\theta/2 = 2m_p\theta$$

Rearranging

$$W_2 + W_1 L_2/(2H) = 2m_p/H \qquad (2)$$

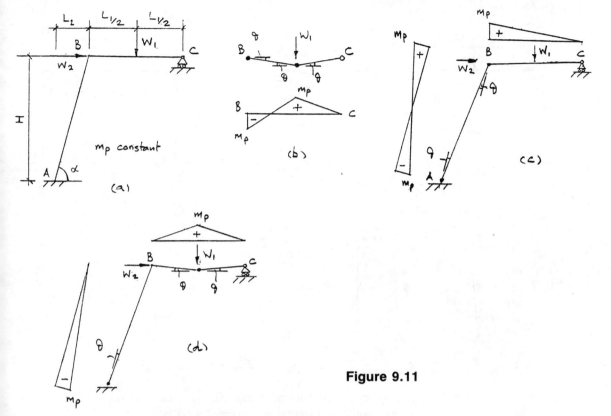

Figure 9.11

For both spans with cancellation of hinges at B (see mechanism shown in Fig. 9.11(d)), equating external and internal work

$$W_1(L_1\theta/2 + L_2\theta/2) + W_2H\theta = 2m_p\theta + m_p\theta$$

Rearranging

$$W_1(1 + L_2/L_1) + 2W_2H/L_1 = 6m_p/L_1 \qquad (3)$$

The bending moment diagrams are shown in Fig. 9.11.
From (1) $20\lambda = 6 \times 60/4$; hence $\lambda = 4.5$
From (2) $\lambda(10 + 20 \times 4/8) = 2 \times 60/4$; hence $\lambda = 1.5$
From (3) $\lambda(20 \times 2 + 2 \times 10 \times 1) = 6 \times 60/4$; hence $\lambda = 1.5$
The lowest load factor $\lambda = 1.5$ and the critical collapse mechanism is either Fig. 9.11(c) or (d).

9.12 Rectangular sway portal frame

Determine the loads at collapse for the portal frame shown in Fig. 9.12(a).

Solution A collapse mechanism may form in beam BC, or a column sway mechanism, or there may be a combined mechanism and a cancellation of hinges at B.

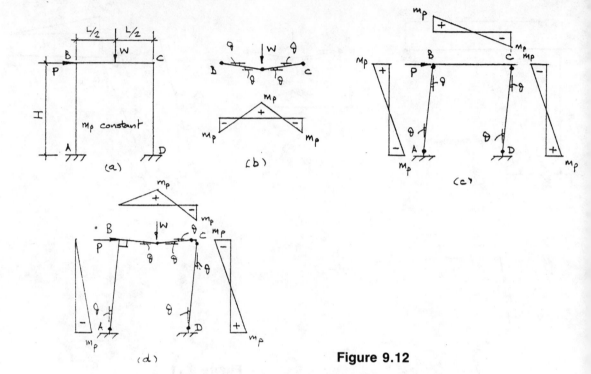

Figure 9.12

For the beam mechanism in member BC (see mechanism shown in Fig. 9.12(b)), equating external and internal work

$$WL\theta/2 = 4m_p\theta; \text{ hence } W = 8m_p/L \tag{1}$$

For the column sway mechanism (see mechanism shown in Fig. 9.12(c)), equating external and internal work

$$PH\theta = 4m_p\theta; \text{ hence } P = 4m_p/H \tag{2}$$

For the combined mechanism and the cancellation of hinges at B (see mechanism shown in Fig. 9.12(d)), equating external and internal work

$$WL\theta/2 + PH\theta = 6m_p\theta$$

Rearranging

$$W + 2PH/L = 12m_p/L \tag{3}$$

The bending moment diagrams are shown in Fig. 9.12.

According to the formula ($n = r + 1$), the number of plastic hinges should be $n = 3 + 1 = 4$. There are three for the beam mechanism which is an exception and four for the other two cases which is correct.

9.13 Portal frame with varying plastic section modulus

Determine the loads at collapse for the portal frame shown in Fig. 9.13(a).

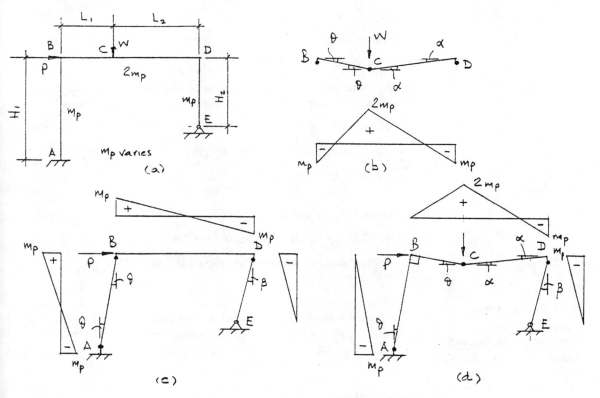

Figure 9.13

Solution   A collapse mechanism may form in beam BD, or a column sway mechanism, or there may be a combined mechanism and a cancellation of hinges at B.

For the beam mechanism in member BC (see mechanism shown in Fig. 9.13(b)), equating external and internal work

$$WL_1\theta = m_p\theta + 2m_p(\theta+\alpha) + m_p\alpha \tag{1}$$

Note that the plastic hinges form in the columns because the plastic moment of resistance in the columns is less than the beam.

From the geometry of the collapsed beam

$$L_1\theta = L_2\alpha \tag{2}$$

Combining (1) and (2)

$$W = (m_p/L_1)(3 + 2L_1/L_2) \tag{3}$$

For the column sway mechanism (see mechanism shown in Fig. 9.13(c)), equating external and internal work

$$PH_1\theta = 2m_p\theta + m_p\beta \tag{4}$$

From the geometry of the collapsed frame

$$H_1\theta = H_2\beta \tag{5}$$

Combining (4) and (5)

$$P = (m_p/H_1)(2+H_1/H_2) \tag{6}$$

For the combined mechanism and the cancellation of hinges at B (see mechanism shown in Fig. 9.13(d)), equating external and internal work

$$WL_1\theta/2+PH_1\theta = m_p\theta+2m_p(\theta+\alpha)+m_p(\beta+\alpha) \tag{7}$$

From the geometry of the collapsed frame

$$L_1\theta = L_2\alpha \tag{8}$$
$$H_1\theta = H_2\beta \tag{9}$$

Combining (7), (8) and (9)

$$W+PH_1/L_1 = (m_p/L_1)(3+3L_1/L_1+H_1/H_1)$$

The bending moment diagrams are shown in Fig. 9.13.

9.14 Portal frame supporting a uniformly distributed load

Determine the loads at collapse for the portal frame shown in Fig. 9.14(a). If $W = wL$ and $L = 2H$ determine the critical collapse load.

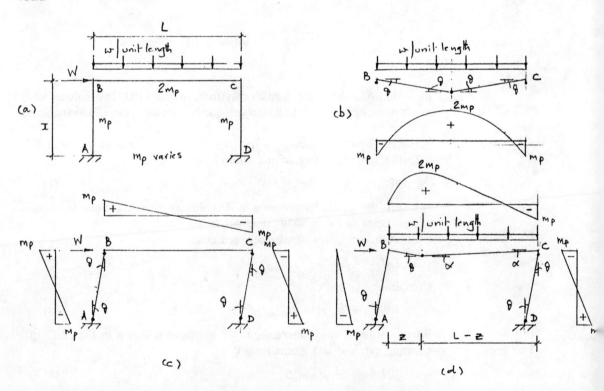

Figure 9.14

Solution A collapse mechanism may form in beam BC, or a column sway mechanism, or there may be a combined mechanism and a cancellation of hinges at B.

For the beam mechanism in member BC (see mechanism shown in Fig. 9.14(b)), equating external and internal work

$$(wL)L\theta/4 = m_p\theta + 2m_p(2\theta) + m_p\theta; \text{ hence } wL = 24m_p/L \qquad (1)$$

Note that the plastic hinges form in the columns because the plastic moment of resistance in the columns is less than the beams.

For the column sway mechanism (see mechanism shown in Fig.9.13(c)), equating external and internal work

$$WH\theta = 4m_p\theta; \text{ hence } W = 4m_p/H \qquad (2)$$

For the combined mechanism and the cancellation of hinges at B (see mechanism shown in Fig. 9.13(d)), the position of the plastic hinge in the beam BC is at a distance z from B. Equating external and internal work

$$WH\theta + wz^2\theta/2 + w(L-z)^2\alpha/2 = m_p\theta + 2m_p(\theta + \alpha) + m_p(\theta + \alpha) + m_p\theta \qquad (3)$$

From the geometry of the collapsed beam BC

$$\theta z = \alpha(L - z) \qquad (4)$$

Combining (3) and (4) and if $W = wL$ and $L = 2H$

$$W = (2m_p/L)(5 - 2z/L)/[1 - (z/L)^2] \qquad (5)$$

W is a minimum when $dW/dz = 0$. Differentiating

$$(5 - 2z/L)/[1 - (z/L)^2] = (-2/L)/(-2z/L^2)$$

Hence

$$z/L = (5 - \sqrt{21})/2 = 0.2087 \qquad (6)$$

Inserting (6) in (5)

$$W = (2m_p/L)2/(5 - \sqrt{21}) = 9.583m_p/L \qquad (7)$$

Comparing the magnitudes of W in (1), (2) and (7) the critical collapse mechanism is sidesway. From (2), $W = 8m_p/L$.

The bending moment diagrams are shown in Fig. 9.14.

9.15 Non-rectangular portal frame with varying plastic section modulus

Determine the loads at collapse for the frame shown in Fig. 9.15(a).

Solution A collapsing mechanism may form in beam BC, or a column sway mechanism, or there may be a combined mechanism and a cancellation of hinges at B.

For the beam mechanism in member BC (see mechanism shown in Fig. 9.15(b)), equating external and internal work

$$W\cos\alpha(L/\cos\alpha)\theta = 7m_p\theta; \text{ hence } W = 7m_p/L \qquad (1)$$

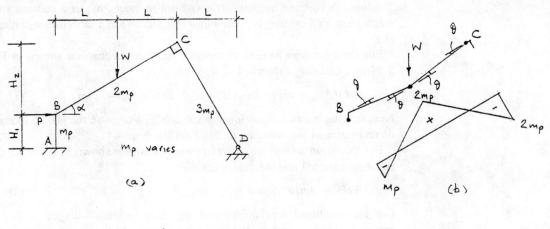

(a)

(b)

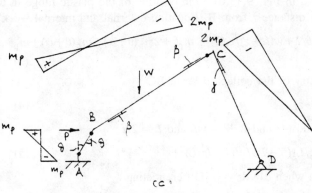

(c)

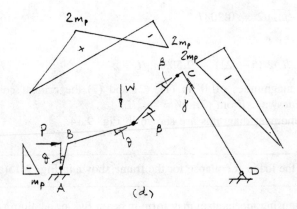

(d)

Figure 9.15

For the column sway mechanism (see mechanism shown in Fig. 9.15(c)), equating external and internal work

$$PH_1\theta + WL\beta = m_p\theta + m_p(\theta+\beta) + 2m_p(\beta+\gamma) \tag{2}$$

From the geometry of the collapsed frame

$$H_1\theta = H_2\beta = (H_1+H_2)\gamma \tag{3}$$

Combining (2) and (3)

$$P+WL/H_2 = (m_p/H_1)[2+3H_1/H_2+2H_1/(H_1+H_2)] \tag{4}$$

For the combined mechanism and the cancellation of hinges at B (see mechanism shown in Fig. 9.15(d)), equating external and internal work

$$WL\gamma+PH_1\theta = 2m_p(\theta+\gamma) \tag{5}$$

From the geometry of the collapsed frame

$$H_1\theta = H_2\beta = (H_1+H_2)\gamma \tag{6}$$

Combining (5) and (6)

$$P+2WL/H_1 = (m_p/H_1)[3+4H_1/H_2+2H_1/(H_1+H_2)]$$

The bending moment diagrams are shown in Fig. 9.15.

9.16 Pitched roof portal frame

Determine the loads at collapse for the portal frame shown in Fig. 9.16(a).

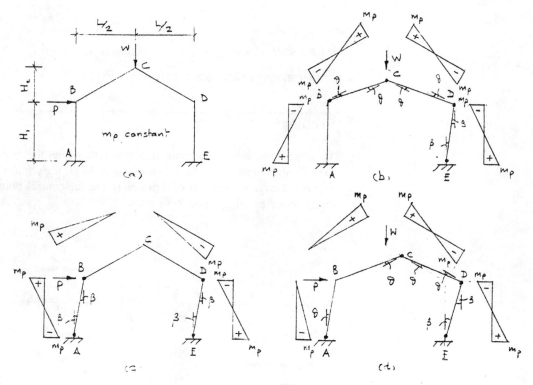

Figure 9.16

Solution A collapse mechanism may form in the pitched roof BCD, or a column sway mechanism, or there may be a combined mechanism and a cancellation of hinges at B.

For the pitched roof mechanism in members BC, CD and CE (see mechanism shown in Fig. 9.16(b)), equating external and internal work

$$WL\theta/2 = 4m_p\theta + 2m_p\beta \tag{1}$$

From the geometry of the collapsed roof

$$H_1\beta = 2H_2\theta \tag{2}$$

Combining (1) and (2)

$$W = (8m_p/L)(1 + H_2/H_1) \tag{3}$$

For the column sway mechanism (see mechanism shown in Fig. 9.16(c)), equating external and internal work

$$PH_1\beta = 4m_p\beta; \text{ hence } P = 4m_p/H_1 \tag{4}$$

For the combined mechanism and the cancellation of hinges at B (see mechanism shown in Fig. 9.16(d)), equating external and internal work

$$WL\theta/2 + PH_1\theta = 3m_p\theta + m_p(\theta + \beta) + m_p\beta \tag{5}$$

From the geometry of the collapsed beam

$$H_1\theta + 2H_2\theta = H_1\beta \tag{6}$$

Combining (5) and (6)

$$P + WL/(2H_1) = (m_p/H_1)[4 + 2(1 + 2H_2/H_1)]$$

The bending moment diagrams are shown in Fig. 9.16.

Problems

1 Determine the plastic moment of resistance about the $x-x$ and $y-y$ axes for the channel section shown in Fig. 9.17. If the section is simply supported over a span AB of 2 m determine the central point load at collapse for bending about both axes.

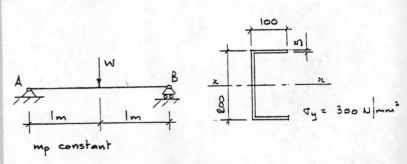

Figure 9.17

Plastic moment of resistance about the $x-x$ axis
$$m_{px} = 2 \times (100 \times 5 \times 97.5 + 95^2 \times 5/2) \times 300 = 42.79\text{E}6 \text{ Nmm}$$
Position of the $y-y$ plastic axis of bending from moments of area
$$200x_p = 2 \times 5 \times (100-x_p) + 190 \times (5-x_p), \text{ hence } x_p = 4.875 \text{ mm}$$
from the back of the channel.
Plastic moment of resistance about the $y-y$ axis
$$m_{py} = (200 \times 4.875^2/2) + 2 \times 5 \times 95.125^2/2 + 190 \times 0.125^2/2) \times 300$$
$$= 14.29\text{E}6 \text{ Nmm}.$$
From $m_p = WL/4$, $W_{px} = 85.575 \text{ kN}$ and $W_{py} = 28.57 \text{ kN}$.

2 Determine the position of the plastic hinge at collapse for the simply supported beam AB which carries a uniformly distributed load of w per unit length over half the span as shown in Fig. 9.18. If the section of the beam is a tube, as shown, and the span is 2 m determine the value of w.

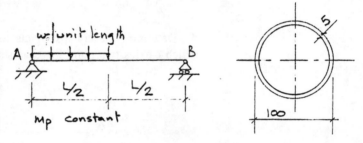

Figure 9.18

Solution

Form the work equation from
$$wz^2\theta_1/2 + w(L-z)^2\theta_2/2 - w(L/2)^2\theta_2/2 = m_p(\theta_1 + \theta_2) \quad \text{and the}$$
geometrical relationship $z\theta_1 = (L-z)\theta_2$. Combine to form
$$w = 2m_pL/[3L^2/(4z) - Lz^2].$$
Equating $\partial w/\partial z = 0$ and solving, $z = 3L/8$ from A.
Substituting back in the work equation, $m_p = 9wL^2/128$.
From the geometry of the cross section
$$m_p = (D_1^3 - D_2^3)\sigma_y/6 = 13.55\text{E}6 \text{ Nmm}.$$
Equating the values of m_p
$$w = 48.18 \text{ kN/m}.$$

3 A beam of constant breadth (B) and variable depth (D) is simply supported over a span (L) as shown in Fig. 9.19. If the beam is loaded with a uniformly distributed load (w) determine the position of the plastic hinge and hence the loading per unit length at collapse.

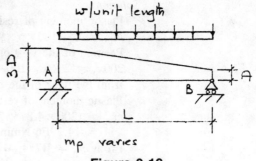

Figure 9.19

Solution

Form the work equation from
$wz^2\theta_2/2 + w(L-z)^2\theta_1/2 = m_p(\theta_1+\theta_2)$, the geometrical relationship
$z\theta_2 = (L-z)\theta_1$, and the section property $z_p = B(D+2zD/L)^2/4$.
Equating $\partial w/\partial z = 0$ and solving, $z/L = (\sqrt{3}-1)/2$ from B.
Substituting z back in the work equation, $w = 6.456BD^2\sigma_y/L^2$.

4 Determine the collapse loads for the simply supported beam AB
which carries two point loads as shown in Fig. 9.20.

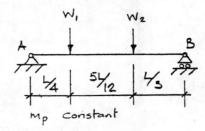

Figure 9.20

Solution

Plastic hinge at W_1.
Form the work equation, $W_1(L/4)\theta_1 + W_2(L/3)\theta_2 = m_p(\theta_1+\theta_2)$ and
the geometrical equation $(L/4)\theta_1 = (L-L/4)\theta_2$. Combining,
$3W_1/4 + W_2/3 = 4m_p/L$.
Plastic hinge at W_2.
Form the work equation, $W_1(L/4)\theta_1 + W_2(L/3)\theta_2 = m_p(\theta_1+\theta_2)$ and
the geometrical equation $(L/3)\theta_2 = (L-L/3)\theta_1$. Combining,
$W_1/4 + 2W_2/3 = 3m_p/L$.

5 Determine the collapse loads for the continuous beam ABCD with
a cantilever shown in Fig. 9.21. Also determine the condition for
simultaneous collapse of span AB and cantilever CD if $wL_1 = W$.

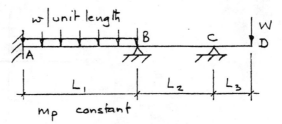

Figure 9.21

Solution

Collapse of span AB.

Form work equation, $wL_1^2\theta/4 = 4m_p\theta$, hence $wL_1 = 16m_p/L_1$.

Collapse of cantilever CD.

From work equation, $WL_3\theta = m_p\theta$, hence $W = m_p/L_3$.

If $wL_1 = W$ then for simultaneous collapse, $L_1/L_3 = 16$.

6 Determine the collapse loads for the continuous beam ABC with two opposing point loads shown in Fig. 9.22.

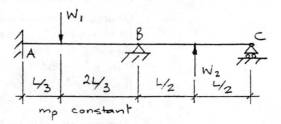

Figure 9.22

Solution

Collapse of span AB.

Form the work equation, $W_1(2L/3)\theta_2 = 2m_p\theta_1 + 2m_p\theta_2$, and the geometrical equation, $(L/3)\theta_1 = (2L/3)\theta_2$. Solving $W_1 = 9m_p/L$.

Collapse of span BC.

Form the work equation $W_2(L/2)\theta = 3m_p\theta$, hence $W_2 = 6m_p/L$.

Combined mechanism.

Form the work equation, $W_1(2L/3)\theta_2 + W_2(L/2)\theta_2 = 3m_p\theta_2 + 2m_p\theta_1$ and combine with the previous geometrical equation to produce $2W_1/3 + W_2/2 = 7m_p/L$.

7 Determine the collapse loads for the rectangular sway frame ABC shown in Fig. 9.23. If $W_1 = 2W_2$ and $H = L$ determine the critical collapse mechanism.

Solution

Collapse of beam BC.

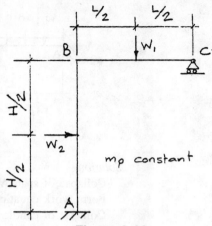

Figure 9.23

Form work equation, $W_1(L/2)\theta = 3m_p\theta$, hence $W_1 = 6m_p/L$.
Collapse of column AB.
Form the work equation, $W_2(H/2)\theta = 4m_p\theta$, hence $W_2 = 8m_p/H$.

Column sway mechanism.
Form the work equation, $W_2(H/2)\theta = 2m_p\theta$, hence $W_2 = 4m_p/H$.

Combined beam and column mechanism.
Form the work equation, $W_2(H/2)\theta + W_1(L/2)\theta = 3m_p\theta$, hence
$W_2 + W_1 L/H = 6m_p/H$.
If $W_1 = 2W_2$ and $H = L$ then the combined mechanism is critical and
$W_1 = 4m_p/L$.

8 Determine the collapse loads for the rectangular portal frame ABCD
shown in Fig. 9.24. If the service loads are $P = 10$ kN and $W = 20$ kN,
$L = H = 5$ m, and $m = 60$ kNm determine the minimum load factor for
collapse.

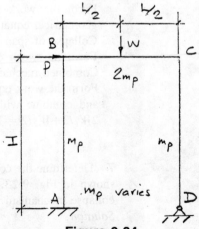

Figure 9.24

Solution

Collapse of beam BC.

From work equation, $W(L/2)\theta = 6m_p\theta$, hence $W = 12m_p/L$.

Column sway mechanism.

From work equation, $PH\theta = 3m_p\theta$, hence $P = 3m_p/H$.

Combined beam and column mechanism.

From work equation, $W(L/2)\theta + PH\theta = 7m_p\theta$, hence $W + 2PH/L = 14m_p/L$.

If numerical values are inserted in equations then the combined beam and column mechanism is critical, and $\lambda = 1.2$.

9 Determine the collapse loads for the unsymmetrical rectangular sway frame ABCD shown in Fig. 9.25.

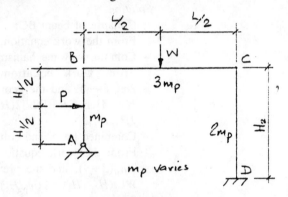

Figure 9.25

Solution

Collapse of beam BC.

From work equation, $W(L/2)\theta = 9m_p\theta$, hence $W_1 = 18m_p/L$.

Collapse of column AB.

From work equation, $P(H_1/2)\theta = 3m_p\theta$, hence $P = 6m_p/H_1$.

Column sway mechanism.

Form the work equation, $P(H_1/2)\theta_1 = m_p\theta_1 + 4m\theta_2$, and the geometrical relationship $H_1\theta_1 = H_2\theta_2$. Combine to form $P = (2m_p/H_1)(1 + 4H_1/H_2)$.

Combined beam and column mechanism.

From the work equation, $W(L/2)\theta_1 + P(H_1/2)\theta_1 = 8m_p\theta_1 + 4m(H_1/H_2)\theta_2$, and the geometrical relationship above, combine to form $W + PH_1/L = (2m_p/L)(8 + 4H_1/H_2)$.

10 Determine the collapse loads for the unsymmetrical sway frame ABCD shown in Fig. 9.26.

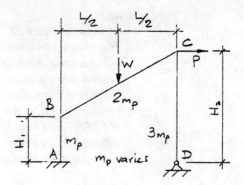

Figure 9.26

Solution

Collapse of beam BC.

From the work equation, $W(L_1/2)\theta = 7m_p\theta$, hence $W = 14m_p/L$.

Column sway mechanism.

From work equation, $PH_2\gamma + W(L/2)\theta = m_p\beta + m_p(\beta+\theta) + 2m_p(\theta+\gamma)$, and the geometrical relationship $H_1\beta = (H_2-H_1)\theta = H_2\gamma$. Hence $P + WL/[2(H_2-H_1)] = (m_p/H_1)[2+3H_1/(H_2-H_1) + 2H_1/H_2]$.

Combined beam and column mechanism.

From the work equation, $PH_2\gamma + W(L/2)2\theta = m_p\theta + 2m_p(\theta+\beta) + 2m_p(\beta+\gamma)$, and the previous geometrical relationship, form $P + WL/(H_2-H_1) = (m_p/H_1)[4+3H_1/(H_2-H_1) + 2H_1/H_2]$.

10

Matrix methods (and BASIC listings)

Introduction and general assumptions

The use of matrices enables generalised algorithms for the analysis of structures to be developed. To an extent the solution of structural problems then becomes a purely mechanical process, although the volume of arithmetic involved is invariably greater than with traditional hand methods and is of a type which does not lend itself readily to hand calculation. Matrix methods are therefore ideally suited to analysis by computer but are not really practicable as hand methods except for academic exercises involving very small structures.

Both the flexibility and stiffness methods can be presented in matrix form, but the stiffness method allows of a much greater degree of generalisation and is the basis for almost all commercial computer programs. BASIC listings of computer programs for both methods are given at the end of this chapter.

The basic assumptions of linearity and small deflections apply, as in hand methods of analysis. In addition it is assumed that structures are composed of straight, uniform members, meeting at joints. In this context a joint is defined as any point at which there is a change in the direction or section properties of a member, or at which a load is applied, or at which two or more members meet. Throughout this chapter letters symbolising matrices are in **bold face type**.

Flexibility method

The flexibility method is a generalisation of the unit load method described in Chapter 7. The loads acting on the structure are expressed by the column vector **W** which contains the external forces and moments at the joints, including dummy zero loads where additional displacements are required, and also the unknown restraints exerted by redundant supports and members. The corresponding displacements vectorially equivalent to **W** are expressed by the vector **Y**. These include external displacements at the joints, settlement at redundant supports, and displacements corresponding to the restraints exerted by redundant members.

As in hand methods of analysis members are assumed to be either pin-jointed and subjected to purely axial forces, or rigidly jointed where flexural effects

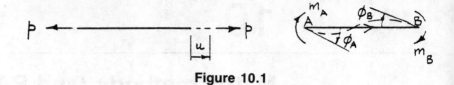

Figure 10.1

predominate and axial displacements can be ignored. Both types of member can appear, however, in the same structure. Since the external loads are applied only at the joints, member forces and displacements relate to the ends of the members, as in Fig. 10.1. Forces and displacements in the members, *including any redundant members*, can therefore be expressed by the column vectors

$$\mathbf{w} = \{p_1, p_2, \text{etc.}, \dots, (m_A, m_B)_1, (m_A, m_B)_2, \text{etc.} \dots\}$$
$$\mathbf{y} = \{u_1, u_2, \text{etc.}, \dots, (\varphi_A, \varphi_B)_1, (\varphi_A, \varphi_B)_2, \text{etc.} \dots\}$$

In an axially loaded member displacement and force are related by the flexibility f, thus

$$u = fp \tag{10.1}$$

where $f = L/EA$. For a structure containing m members $\tag{10.2}$

$$
\begin{bmatrix} u_1 \\ u_2 \\ \cdot \\ \cdot \\ \cdot \\ u_m \end{bmatrix}
=
\begin{bmatrix} f_1 & & & \\ & f_2 & & \\ & & \cdot & \\ & & & \cdot \\ & & & & f_m \end{bmatrix}
\begin{bmatrix} p_1 \\ p_2 \\ \cdot \\ \cdot \\ \cdot \\ p_m \end{bmatrix}
\tag{10.3}
$$

A similar relationship for a member in flexure may be obtained from the *slope-deflection* equations (4.1) and (4.2), which may be written in matrix form as follows

$$
\begin{bmatrix} \varphi_A \\ \varphi_A \end{bmatrix}
= f
\begin{bmatrix} 2 & -1 \\ -1 & 2 \end{bmatrix}
\begin{bmatrix} m_A \\ m_B \end{bmatrix}
\tag{10.4}
$$

where $f = L/6EI$ in this case. Hence, for m members $\tag{10.5}$

$$
\begin{bmatrix} \varphi_{A1} \\ \varphi_{B1} \\ \varphi_{A2} \\ \varphi_{B2} \\ \cdot \\ \cdot \\ \varphi_{Am} \\ \varphi_{Bm} \end{bmatrix}
=
\begin{bmatrix} f_1 \begin{bmatrix} 2 & -1 \\ -1 & 2 \end{bmatrix} & & & \\ & f_2 \begin{bmatrix} 2 & -1 \\ -1 & 2 \end{bmatrix} & & \\ & & \cdot & \\ & & & f_m \begin{bmatrix} 2 & -1 \\ -1 & 2 \end{bmatrix} \end{bmatrix}
\begin{bmatrix} m_{A1} \\ m_{B1} \\ m_{A2} \\ m_{B2} \\ \cdot \\ \cdot \\ m_{Am} \\ m_{Bm} \end{bmatrix}
\tag{10.6}
$$

Combinations of equations (10.3) and (10.6) are also possible where both kinds of member are present in the same structure. In general these equations of elasticity may be written as

$$\mathbf{y} = \mathbf{fw} \tag{10.7}$$

where **f** is the *flexibility matrix of the members*. It is a square matrix with either single elements or submatrices on the leading diagonal. The remainder of the elements are zeros.

The internal and external forces are related by equilibrium equations which may be written generally as

$$\mathbf{w} = \mathbf{BW} \tag{10.8}$$

where **B** is a matrix of influence coefficients.

The elements of **B** are determined, a column at a time, by successive analyses in which the load vector contains a single unit load. Typically the ith column of **B** consists of the member forces produced when the ith element of **W** is unity and all the other elements are zero. Since the load vector contains all the redundant restraints, and on each analysis all the loads except one are zero, the structure is effectively statically determinate at this stage.

In the next stage the principle of virtual work is used to generate equations of compatibility between the external displacements at the loaded joints and the internal displacements in the members, i.e. between vectors **Y** and **y**. Each equation is generated by the corresponding unit virtual load in **W** and the column of **B** in equilibrium with it.

For example, in a pin-jointed frame let the deflection Y_i at one of the external joints be the ith element in the displacement vector **Y**. The external deflections are compatible with the actual member extensions **y** due to the external loads and redundant restraints, i.e. $u_1, u_2, \ldots, u_m$. Now let the ith element of the load vector **W** be a unit virtual load vectorially equivalent to Y_i, all other elements being zero. The internal member forces in equilibrium with this load are the influence coefficients in the ith column of **B**. Let these be $p_1, p_2, \ldots, p_m$. The virtual work equation (1.5) then becomes

$$1 \times Y_i = \Sigma(p_1 u_1 + p_2 u_2 + \ldots + p_m u_m) \tag{10.9}$$

The complete set of compatibility equations may therefore be written as

$$\mathbf{Y} = \mathbf{B'y} \tag{10.10}$$

This is a *contragredient transformation* of equation (10.8). It is the general form of the transformation achieved in Chapter 8 Problem 6 by applying the Müller Breslau principle.

By combining equations (10.7) and (10.8) and then substituting for **y** in equation (10.10)

$$\mathbf{Y} = \mathbf{B'fBW} \tag{10.11}$$

which may be written as

$$\mathbf{Y} = \mathbf{FW} \tag{10.12}$$

where $\mathbf{F} = \mathbf{B'fB}$ and is the flexibility matrix of the assembled structure. It is a square matrix of flexibility coefficients, symmetrical about the leading diagonal.

When the unit virtual load corresponds with the restraint exerted by a redundant member there are no external loads, so the left hand side of equation

(10.9) is zero. Hence the displacements in equations (10.11) corresponding to redundant restraints are either zero, in the case of redundant members, or values of settlement, assumed to be known, at redundant supports. These equations can therefore be solved to obtain the redundant restraints. The displacements at the loaded joints can then be obtained directly from the other equations in (10.11) and the member forces from equations (10.8).

The effects of lack of fit and thermal strains are to modify the member displacements. Equation (10.11) then becomes

$$Y = B'(fBW + \lambda_T + \lambda_F) \tag{10.13}$$

where λ_T and λ_F are vectors of member displacements due to thermal expansion and lack of fit respectively. These equations can be divided into three groups by ordering the elements of Y and W and partitioning matrix B so that the groups represent external joints, redundant supports, and redundant members respectively as follows

$$B = |B_W \; B_{RS} \; B_{RM}| \tag{10.14}$$

Transposing and substituting for B' in equation (10.13) then gives

$$Y_W = B'_W(fBW + \lambda_T + \lambda_F) \tag{10.15}$$
$$Y_S = B'_{RS}(fBW + \lambda_T + \lambda_F) \tag{10.16}$$
$$0 = B'_{RM}(fBW + \lambda_T + \lambda_F) \tag{10.17}$$

where Y_W and Y_S are the displacements in the directions of the external loads and redundant reactions respectively and 0 is a null vector.

If it is assumed for convenience that the redundant members are the last to be assembled, then the lack of fit for all other members will be zero and hence $B'_w\lambda_F$ and $B'_{RS}\lambda_F$ become null vectors and $B'_{RM}\lambda_F$ consists of the non-zero elements of λ_F. Taking these over to the left hand side in equation (10.17), the three equations can be written as

$$Y = B'(fBW + \lambda_T) \tag{10.18}$$

where the elements of Y for redundant members are the lack of fit with the sign changed, i.e. in a pin-jointed frame they are positive when members are too short. In general the lack-of-fit displacements in Y are positive when the forces required to make the members fit are positive.

Determination of the elements of matrix B is usually carried out by hand because it is not possible, except in the case of certain stereotyped structures, to construct a sufficiently general algorithm for analysis by computer. The remainder of the analysis can, however, be performed entirely by computer and listings of programs for structures with pin-jointed or rigidly jointed members are given at the end of this chapter.

Worked examples

10.1 Deflection of a statically determinate frame

Given that all the members of the pin-jointed frame in Fig. 10.2 have the same sectional area and Young's modulus, obtain expressions for the vertical deflection at A and the horizontal deflections at B and C.

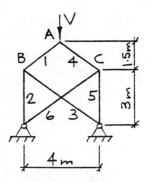

Figure 10.2

Solution (1) Number the members as shown and construct the elasticity equations $\mathbf{y} = \mathbf{fw}$. As all the members have the same value of EA, their flexibilities are in proportion to their lengths. Hence

$$
\begin{bmatrix} u_1 \\ u_2 \\ u_3 \\ u_4 \\ u_5 \\ u_6 \end{bmatrix} = 1/EA \begin{bmatrix} 2.5 & & & & & \\ & 3.0 & & & & \\ & & 5.0 & & & \\ & & & 2.5 & & \\ & & & & 3.0 & \\ & & & & & 5.0 \end{bmatrix} \begin{bmatrix} p_1 \\ p_2 \\ p_3 \\ p_4 \\ p_5 \\ p_6 \end{bmatrix}
$$

(2) Construct the equilibrium equations $\mathbf{w} = \mathbf{BW}$.

The load vector $\mathbf{W}$ consists of the load V at A and dummy zero loads for the deflections at B and C. However, since the structure is symmetrical, only one of the latter is required, say at C. The first column of $\mathbf{B}$ consists of the member forces produced by a single unit downward load at A, the second by a unit horizontal load to the right at C. These may be obtained by any of the usual methods for the analysis of pin-jointed plane frames (see Chapter 7, *Solving Problems in Structures*, Volume 1). Hence

$$
\begin{bmatrix} p_1 \\ p_2 \\ p_3 \\ p_4 \\ p_5 \\ p_6 \end{bmatrix} = \begin{bmatrix} -5/6 & 0 \\ -1 & 0 \\ 5/6 & 0 \\ -5/6 & 0 \\ -1 & -3/4 \\ 5/6 & 5/4 \end{bmatrix} \begin{bmatrix} V \\ 0 \end{bmatrix}
$$

(3) Compute $\mathbf{F} = \mathbf{B'fB}$ to give the equations $\mathbf{Y} = \mathbf{FW}$.

$$
\begin{bmatrix} Y_A \\ Y_C \end{bmatrix} = 1/EA \begin{bmatrix} 16.42 & 7.46 \\ 7.46 & 9.5 \end{bmatrix} \begin{bmatrix} V \\ 0 \end{bmatrix}
$$

(4) As there are no redundancies multiply out to obtain the deflections.

$$
\begin{bmatrix} Y_A \\ Y_C \end{bmatrix} = V/EA \begin{bmatrix} 16.42 \\ 7.46 \end{bmatrix}
$$

An additional member BC, of the same section and material, was added to the frame in the previous problem, but was found to be 4 mm too short on assembly. Given that $A = 594\ \text{mm}^2$ and $E = 200\ \text{kN/mm}^2$, determine

(a) the deflection at joint A and the forces in all the members when the frame is assembled but not loaded (the self-straining forces),

(b) the vertical load that would be required at A if there is to be no deflection there.

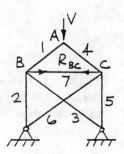

Figure 10.3

Solution − *Fig. 10.3* (1) Number the members as shown and construct $\mathbf{y} = \mathbf{fw}$. This is the same as in the previous problem, with the addition of the redundant member 7 (BC).

$$\begin{bmatrix} u_1 \\ u_2 \\ u_3 \\ u_4 \\ u_5 \\ u_6 \\ u_7 \end{bmatrix} = 1/EA \begin{bmatrix} 2.5 & & & & & & \\ & 3.0 & & & & & \\ & & 5.0 & & & & \\ & & & 2.5 & & & \\ & & & & 3.0 & & \\ & & & & & 5.0 & \\ & & & & & & 4.0 \end{bmatrix} \begin{bmatrix} p_1 \\ p_2 \\ p_3 \\ p_4 \\ p_5 \\ p_6 \\ p_7 \end{bmatrix}$$

(2) Construct $\mathbf{w} = \mathbf{BW}$, noting that the force in member 7 appears as p_7 in $\mathbf{w}$ and as R_{BC} in $\mathbf{W}$. The first column of $\mathbf{B}$ is the same as in the previous problem, but with the redundant member force p_7 equal to zero. The second column is produced by a unit tensile force in member 7.

$$\begin{bmatrix} p_1 \\ p_2 \\ p_3 \\ p_4 \\ p_5 \\ p_6 \\ p_7 \end{bmatrix} = \begin{bmatrix} -5/6 & 0 \\ -1 & 3/4 \\ 5/6 & -5/4 \\ -5/6 & 0 \\ -1 & 3/4 \\ 5/6 & -5/4 \\ 0 & 1 \end{bmatrix} \begin{bmatrix} V \\ R_{BC} \end{bmatrix}$$

(3) Compute $\mathbf{F} = \mathbf{B'fB}$ to give the equations $\mathbf{Y} = \mathbf{FW}$. For part (a) of the problem $V = 0$ and $Y_{BC} = 0.004\ \text{m}$, the lack of fit. *Units must be in m to be consistent with f and the sign is positive as the member is too short.*

$$\begin{bmatrix} Y_A \\ 0.004 \end{bmatrix} = 1/EA \begin{bmatrix} 16.42 & -14.92 \\ -14.92 & 23.00 \end{bmatrix} \begin{bmatrix} 0 \\ R_{BC} \end{bmatrix}$$

(4) Determine the redundant restraint R_{BC}. Multiplying out, the second equation becomes

$$0.004 = 23.00 R_{BC}/EA$$

and $EA = 200 \times 594 = 118.8E3$ kN
Hence

$$R_{BC} = 20.66 \text{ kN}$$

(5) Back-substitute for R_{BC} to obtain Y_A.

$$Y_A = -14.92 R_{BC}/EA = -2.60E\text{-}3 \text{ m (i.e. upwards)}$$

(6) Back-substitute in $\mathbf{w} = \mathbf{BW}$ to obtain the member forces.

$$\begin{bmatrix} p_1 \\ p_2 \\ p_3 \\ p_4 \\ p_5 \\ p_6 \\ p_7 \end{bmatrix} = \begin{bmatrix} 0 \\ 15.50 \\ -25.83 \\ 0 \\ 15.50 \\ -25.83 \\ 20.66 \end{bmatrix} \text{ kN}$$

Solution In the equation $\mathbf{Y} = \mathbf{FW}$, $Y_A = 0$ and both the loads V and R_{BC} are now unknown. The lack of fit and the flexibility matrix are unchanged. Solving, $V = 45.73$ kN, $R_{BC} = 50.32$ kN.

10.3 Thermal strains

In the pin-jointed frame of Fig. 10.4(a) all the members are of steel, except for AB and BC which are of aluminium alloy. Sectional and material properties are:

	Area mm^2	Young's modulus kN/mm^2	Linear expansion coeff. /°C
Steel	2000	200	12E-6
Alloy	4000	70	20E-6

If there is no lack of fit on assembly determine the forces in the members and the vertical deflection of joint D:
  (a) when loaded immediately after assembly,
  (b) after a rise in temperature of 15°C.

Solution (1) Construct $\mathbf{y} = \mathbf{fw}$ by determining L/EA for each member. The units are mm/kN.

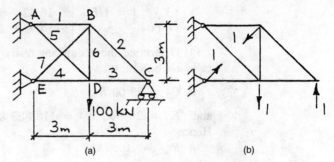

Figure 10.4

$$\begin{bmatrix} u_1 \\ u_2 \\ u_3 \\ u_4 \\ u_5 \\ u_6 \\ u_7 \end{bmatrix} = 1/1000 \begin{bmatrix} 10.71 & & & & & & \\ & 15.15 & & & & & \\ & & 7.5 & & & & \\ & & & 7.5 & & & \\ & & & & 10.61 & & \\ & & & & & 7.5 & \\ & & & & & & 10.61 \end{bmatrix} \begin{bmatrix} p_1 \\ p_2 \\ p_3 \\ p_4 \\ p_5 \\ p_6 \\ p_7 \end{bmatrix}$$

(2) Construct $\mathbf{w} = \mathbf{BW}$. Selecting member BE and the support at C as redundancies the columns of $\mathbf{B}$ are obtained by applying each of the three unit load cases shown in Fig. 10.4(b) in turn.

$$\begin{bmatrix} p_1 \\ p_2 \\ p_3 \\ p_4 \\ p_5 \\ p_6 \\ p_7 \end{bmatrix} = \begin{bmatrix} 0 & -1 & -0.7071 \\ 0 & -1.414 & 0 \\ 0 & 1 & 0 \\ -1 & 2 & -0.7071 \\ 1.414 & -1.414 & 1 \\ 0 & 1 & -0.7071 \\ 0 & 0 & 1 \end{bmatrix} \begin{bmatrix} 100 \\ R_C \\ R_{BE} \end{bmatrix}$$

(3) Compute $\mathbf{F} = \mathbf{B}'\mathbf{fB}$ and construct $\mathbf{Y} = \mathbf{FW}$. Initially there is no lack of fit or settlement. *This and the next two stages of the problem can be carried out either by hand or by means of the computer program at the end of the chapter.*

$$\begin{bmatrix} Y_D \\ 0 \\ 0 \end{bmatrix} = 1/1000 \begin{bmatrix} 28.71 & -36.21 & 20.31 \\ -36.21 & 107.20 & -23.34 \\ 20.31 & -23.34 & 34.07 \end{bmatrix} \begin{bmatrix} 100 \\ R_C \\ R_{BE} \end{bmatrix}$$

Multiplying out and transposing, the last two equations become

$$107.70 R_C - 23.34 R_{BE} = 3621$$
$$-23.34 R_C + 34.07 R_{BE} = -2031$$

Solving, $R_C = 24.45$ kN, $R_{BE} = -42.85$ kN

(4) Back-substitute in $\mathbf{Y} = \mathbf{FW}$ to obtain Y_D.

$$Y_D = (2871 - 36.21 \times 24.45 - 20.31 \times 42.85)/1000 = 1.1 \text{ mm}$$

(5) Back-substitution of R_C and R_{BE} in $\mathbf{w} = \mathbf{BW}$ gives the member forces

$$\begin{bmatrix} p_1 \\ p_2 \\ p_3 \\ p_4 \\ p_5 \\ p_6 \\ p_7 \end{bmatrix} = \begin{bmatrix} 5.85 \\ -34.57 \\ 24.45 \\ -20.80 \\ 63.98 \\ 54.75 \\ -42.85 \end{bmatrix}$$

Solution (1) Calculate in mm the additional extensions of the members due to the rise in temperature using $L\alpha t$.

$$\lambda_T = \begin{bmatrix} 0.900 \\ 1.273 \\ 0.540 \\ 0.540 \\ 0.764 \\ 0.540 \\ 0.764 \end{bmatrix}$$

(2) Compute $\mathbf{B}'\lambda_T$; then solve equation (10.18), i.e. $\mathbf{B}'\mathbf{fBW} + \mathbf{B}'\lambda_T$, noting that $\mathbf{B}'\mathbf{fBW}$ remains unchanged.

$$\begin{bmatrix} Y_D \\ 0 \\ 0 \end{bmatrix} = 1/1000 \begin{bmatrix} 28.71 & -36.21 & 20.31 \\ -36.21 & 107.20 & -23.34 \\ 20.31 & -23.34 & 34.07 \end{bmatrix} \begin{bmatrix} 100 \\ R_C \\ R_{BE} \end{bmatrix} + \begin{bmatrix} 0.540 \\ -1.620 \\ 0.127 \end{bmatrix}$$

Solving the last two equations gives $R_C = 7.64\,\text{kN}$, $R_{BE} = -50.65\,\text{kN}$ and hence $Y_D = 2.1\,\text{mm}$.

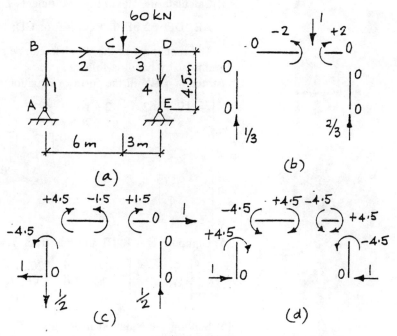

Figure 10.5

10.4 Rigidly jointed frame

Given that the rigidity EI of the members of the frame in Fig. 10.5(a) are: beams $25E3 \text{ kNm}^2$, columns $12E3 \text{ kNm}^2$, determine
(a) the vertical deflection under the load and the horizontal sway,
(b) the bending moments in the frame.

Solution (1) In order to comply with the assumption of straight members, loaded only at their ends, there must be joints at A,B,C,D, and E. There is only one redundancy. Assume that this is the horizontal reaction at E and make the frame statically determinate by setting this to zero initially. Number the members and indicate ends A and B by orientation arrows, as shown.

(2) Construct $\mathbf{w} = \mathbf{BW}$. The load vector $\mathbf{W}$ consists of the load at C, a dummy horizontal load at D to give the sway, and the redundant restraint at E. The columns of $\mathbf{B}$ are the moments (positive clockwise) at the ends of the members, produced by replacing each element of $\mathbf{W}$ in turn by a single unit force, as shown in Fig. 10.5(b), (c), and (d).

$$
\begin{bmatrix}
m_{A1} \\ m_{B1} \\ m_{A2} \\ m_{B2} \\ m_{A3} \\ m_{B3} \\ m_{A4} \\ m_{B4}
\end{bmatrix}
=
\begin{bmatrix}
0 & 0 & 0 \\
0 & -4.5 & 4.5 \\
0 & 4.5 & -4.5 \\
-2.0 & -1.5 & 4.5 \\
2.0 & 1.5 & -4.5 \\
0 & 0 & 4.5 \\
0 & 0 & -4.5 \\
0 & 0 & 0
\end{bmatrix}
\begin{bmatrix}
60 \\ 0 \\ R_E
\end{bmatrix}
$$

(3) Calculate the flexibility coefficients $L/6EI$ of the members in m/kN.

AB, DE: 62.5E-6; BC: 40E-6; CD: 20E-6

(4) The remainder of the problem can be solved, either with the computer program, or by hand as follows.
Construct $\mathbf{w} = \mathbf{fW}$ in the form of equation (10.6).

$$
\begin{bmatrix}
\varphi_{A1} \\ \varphi_{B1} \\ \varphi_{A2} \\ \varphi_{B2} \\ \varphi_{A3} \\ \varphi_{B3} \\ \varphi_{A4} \\ \varphi_{B4}
\end{bmatrix}
= 1E\text{-}6
\begin{bmatrix}
62.5 \begin{bmatrix} 2 & -1 \\ -1 & 2 \end{bmatrix} & & & \\
& 40.0 \begin{bmatrix} 2 & -1 \\ -1 & 2 \end{bmatrix} & & \\
& & 20.0 \begin{bmatrix} 2 & -1 \\ -1 & 2 \end{bmatrix} & \\
& & & 62.5 \begin{bmatrix} 2 & -1 \\ -1 & 2 \end{bmatrix}
\end{bmatrix}
\begin{bmatrix}
m_{A1} \\ m_{B1} \\ m_{A2} \\ m_{B2} \\ m_{A3} \\ m_{B3} \\ m_{A4} \\ m_{B4}
\end{bmatrix}
$$

(5) Compute $\mathbf{F} = \mathbf{B'fB}$ and construct $\mathbf{Y} = \mathbf{FW}$.

$$
\begin{bmatrix}
Y_C \\ Y_D \\ 0
\end{bmatrix}
=
\begin{bmatrix}
480.0E\text{-}6 & 720.0E\text{-}6 & -1.620E\text{-}3 \\
720.0E\text{-}6 & 4.961E\text{-}3 & -6.176E\text{-}3 \\
-1.620E\text{-}3 & -6.176E\text{-}3 & 12.35E\text{-}3
\end{bmatrix}
\begin{bmatrix}
60 \\ 0 \\ R_E
\end{bmatrix}
$$

(6) Solving the equation in R_E gives $R_E = 7.869 \text{ kN}$.
(7) Back-substitute to obtain $Y_C = 16.05E\text{-}3$ m, $Y_D = -5.40E\text{-}3$ m. The

negative sign of Y_C indicates that the horizontal sway is to the left, i.e. opposite to the direction of the unit load.

(8) Back-substitute in $\mathbf{w} = \mathbf{BW}$ for the member forces.

$$
\begin{bmatrix} m_{A1} \\ m_{B1} \\ m_{A2} \\ m_{B2} \\ m_{A3} \\ m_{B3} \\ m_{A4} \\ m_{B4} \end{bmatrix} = \begin{bmatrix} 0 \\ 35.41 \\ -35.41 \\ -84.59 \\ 84.59 \\ 35.41 \\ -35.41 \\ 0 \end{bmatrix}
$$

(9) Convert to the bending moment convention for frames, i.e. positive curvature is concave.

$$M_A, M_E = 0, \quad M_B, M_D = -35.41 \text{ kNm}, \quad M_C = 84.58 \text{ kNm}$$

Stiffness method

In the stiffness method the analysis starts with the development of compatibility equations relating the member displacements, in local coordinates, with the joint displacements, in the general coordinates of the structure. Consider the pin-jointed frame in Fig. 10.6(a). On the line diagram of the frame members and joints are numbered sequentially (joint numbers in brackets) and each member is arbitrarily given a positive orientation, indicated by an arrow. The orientation serves to define ends A and B of the member and the corresponding joint numbers i and j. At this stage numbers can be considered to be allocated arbitrarily, although it will be shown later that running time and storage in computer analyses are usually reduced when for every member the difference between the joint numbers at its ends is small.

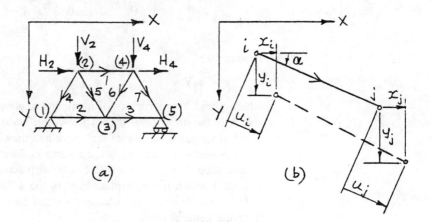

Figure 10.6

Pinned joints

In a plane pin-pointed structure the displacement of a joint is fully defined by the displacements x and y parallel to the general axes X and Y. Joint rotation has no meaning as all the members connected to it are free to rotate relative to one another. Member displacements are axial, positive in the direction of orientation. For a typical member of length L, from Fig. 10.6(b),

$$u_i = lx_i + my_i \tag{10.19}$$
$$u_j = lx_j + my_j \tag{10.20}$$

where l and m are the direction cosines, which may be obtained from the joint coordinates, thus

$$l = \cos\alpha = (X_j - X_i)/L \tag{10.21}$$
$$m = \cos(90 - \alpha) = (Y_j - Y_i)/L \tag{10.22}$$
where $L = [(X_j - X_i)^2 + (Y_j - Y_i)^2]^{1/2}$. $\tag{10.23}$

The extension of the member is therefore given by

$$u = u_j - u_i = -lx_i - my_i + lx_j + my_j \tag{10.24}$$

The complete set of compatibility equations for the frame is obtained by repeating this equation for each member, allocating joint numbers to i and j in accordance with the orientation arrows. For the frame in Fig. 10.6(a), which has seven members, the compatibility equations may be written in matrix form as follows.

$$
\begin{bmatrix} u_1 \\ u_2 \\ u_3 \\ u_4 \\ u_5 \\ u_6 \\ u_7 \end{bmatrix}
=
\overset{\text{joint} \quad\; 2 \qquad\quad 3 \qquad\quad 4 \qquad 5}{
\begin{bmatrix}
-l_1 & -m_1 & 0 & 0 & l_1 & m_1 & 0 \\
0 & 0 & l_2 & m_2 & 0 & 0 & 0 \\
0 & 0 & -l_3 & -m_3 & 0 & 0 & l_3 \\
-l_4 & -m_4 & 0 & 0 & 0 & 0 & 0 \\
-l_5 & -m_5 & l_5 & m_5 & 0 & 0 & 0 \\
0 & 0 & l_6 & m_6 & -l_6 & -m_6 & 0 \\
0 & 0 & 0 & 0 & -l_7 & -m_7 & l_7
\end{bmatrix}}
\begin{bmatrix} x_2 \\ y_2 \\ x_3 \\ y_3 \\ x_4 \\ y_4 \\ x_5 \end{bmatrix}
$$

$$\tag{10.25}$$

In general

$$\mathbf{y} = \mathbf{AY} \tag{10.26}$$

where $\mathbf{y}$ and $\mathbf{Y}$ are vectors of member and joint displacements respectively and $\mathbf{A}$ is the *displacement transformation matrix*.

When joint displacements are restrained they are omitted from $\mathbf{Y}$, and the corresponding columns of $\mathbf{A}$ are absent. In the example the restraints are the pinned support at joint 1 which has no displacements, and the roller support at joint 5 which has a displacement in the x direction only.

The elasticity equations relating the forces and displacements in the members have the general form

$$\mathbf{w} = \mathbf{ky} \tag{10.27}$$

where $\mathbf{k}$ is the stiffness matrix for the members. In the example

$$
\begin{bmatrix} p_1 \\ p_2 \\ p_3 \\ p_4 \\ p_5 \\ p_6 \\ p_7 \end{bmatrix} = \begin{bmatrix} k_1 & & & & & \\ & k_2 & & \text{zeros} & & \\ & & k_3 & & & \\ & & & k_4 & & \\ & & & & k_5 & \\ & \text{zeros} & & & k_6 & \\ & & & & & & k_7 \end{bmatrix} \begin{bmatrix} u_1 \\ u_2 \\ u_3 \\ u_4 \\ u_5 \\ u_6 \\ u_7 \end{bmatrix}
$$

(10.28)

where k_1, k_2, etc. are the axial stiffnesses of the members, given by

$$k = EA/L \tag{10.29}$$

Now, using the principle of contragredience, the compatibility equation (10.26) is transformed into the equilibrium equation

$$\mathbf{W} = \mathbf{A'w} \tag{10.30}$$

and, combining equations (10.27) and (10.26)

$$\mathbf{w} = \mathbf{kAY} \tag{10.31}$$

Hence, substituting for $\mathbf{w}$ in equation (10.30)

$$\mathbf{W} = \mathbf{KY} \tag{10.32}$$

where $\mathbf{K} = \mathbf{A'kA}$ and is the stiffness matrix of the assembled structure, $\mathbf{K}$ is a square, symmetrical matrix, whose order is equal to the number of degrees of freedom of the joints.

Equation (10.32) is a set of simultaneous equations which may be solved to obtain the joint displacements $\mathbf{Y}$. Finally, back-substitution into equation (10.31) yields the member forces.

Worked example

10.5 Basic equations

For the truss in Fig. 10.7 derive the following, given the numbering and orientations shown and assuming that all members have the same section and are of the same material:

(a) equations of compatibility between the member and joint displacements

(b) the matrix of member stiffnesses

(c) a set of equations from which the member forces may be calculated

(d) a set of equations which may be solved for the joint displacements.

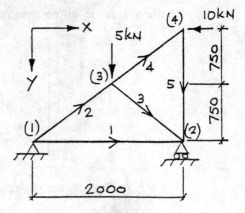

Figure 10.7

Solution (a)

Member	Joints		Length	Direction cosines	
	i	j	L	$l = (X_j - X_i)/L$	$m = (Y_j - Y_i)/L$
1	1	2	2000	1	0
2	1	3	1250	0.8	−0.6
3	3	2	1250	0.8	0.6
4	3	4	1250	0.8	−0.6
5	4	2	1500	0	1

Joint displacement vector $\mathbf{Y} = \{x_2, x_3, y_3, x_4, y_4\}$
Member displacement vector $\mathbf{y} = \{u_1, u_2, u_3, u_4, u_5\}$
Construct the compatibility equations $\mathbf{y} = \mathbf{AY}$, noting that columns corresponding to x_i, y_i contain $-l, -m$, while those corresponding to x_j, y_j contain $+l, +m$.

$$
\begin{bmatrix} u_1 \\ u_2 \\ u_3 \\ u_4 \\ u_5 \end{bmatrix} = \begin{bmatrix} 1 & 0 & 0 & 0 & 0 \\ 0 & 0.8 & -0.6 & 0 & 0 \\ 0.8 & -0.8 & -0.6 & 0 & 0 \\ 0 & -0.8 & 0.6 & 0.8 & -0.6 \\ 0 & 0 & 0 & 0 & -1 \end{bmatrix} \begin{bmatrix} x_2 \\ x_3 \\ y_3 \\ x_4 \\ y_4 \end{bmatrix}
$$

Solution (b) Stiffness $= EA/L$. Since EA is common to all members, their stiffnesses are inversely proportional to their lengths. Taking $EA/1000$ as a common factor

$$
\mathbf{k} = EA/1000 \begin{bmatrix} 0.5 & & & & \\ & 0.8 & & & \\ & & 0.8 & & \\ & & & 0.8 & \\ & & & & 0.667 \end{bmatrix}
$$

Solution (c) Compute $\mathbf{kA}$ and construct the equations $\mathbf{w} = \mathbf{kAY}$

$$
\begin{bmatrix} u_1 \\ u_2 \\ u_3 \\ u_4 \\ u_5 \end{bmatrix} =
\begin{bmatrix}
0.50 & 0 & 0 & 0 & 0 \\
0 & 0.64 & -0.48 & 0 & 0 \\
0.64 & -0.64 & -0.48 & 0 & 0 \\
0 & -0.64 & 0.48 & 0.64 & -0.48 \\
0 & 0 & 0 & 0 & -0.667
\end{bmatrix}
\begin{bmatrix} x_2 \\ x_3 \\ y_3 \\ x_4 \\ y_4 \end{bmatrix}
$$

Solution (d) Compute $\mathbf{K} = \mathbf{A}'k\mathbf{A}$ and construct the equations $\mathbf{W} = \mathbf{KY}$. The load vector $\mathbf{W} = \{0,0,5,-10,0\}$.

$$
\begin{bmatrix} 0 \\ 0 \\ 5 \\ -10 \\ 0 \end{bmatrix} = EA/1000
\begin{bmatrix}
1.012 & -0.512 & -0.384 & 0 & 0 \\
-0.512 & 1.536 & -0.384 & -0.512 & 0.384 \\
-0.384 & -0.384 & 0.864 & 0.384 & -0.288 \\
0 & -0.512 & 0.384 & 0.512 & -0.384 \\
0 & 0.384 & -0.288 & -0.384 & 0.955
\end{bmatrix}
\begin{bmatrix} x_2 \\ x_3 \\ y_3 \\ x_4 \\ y_4 \end{bmatrix}
$$

The symmetry of the stiffness matrix provides a check on the arithmetic.

Rigid joints

The matrix equations derived for structures with pinned joints are equally valid for rigid joints, provided that the contents of matrices are suitably modified.

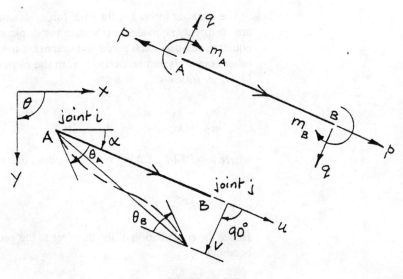

Figure 10.8

Consider a typical member AB of length L, connected between joints i and j, as shown in Fig. 10.8. The member displacements are $\{u,v,\theta_A,\theta_B\}$, where u is the extension, v the sway, and θ_A,θ_B the end rotations. The joint displacements are $\{x_i,y_i,\theta_i \ldots x_j,y_j,\theta_j\}$, where x and y are parallel to the general axes, and θ_i and θ_j are the rotations. The dots indicate that joints i

and j are not necessarily adjacent in the joint displacement vector. The sign convention is in accordance with a right hand set of axes (see *Solving Problems in Structures*, Volume 1, p 152). All the displacements in the figure are therefore positive.

The extension u is given by equation (10.24), and the sway v may be obtained from the same equation by replacing α with $(\alpha+90°)$. The end rotations θ_A and θ_B, since by definition there is no relative rotation of the members connected to a rigid joint, are equal to the joint rotation θ_i and θ_j. Hence the contribution of the member to the compatibility equations $\mathbf{y} = \mathbf{AY}$ is

$$
\begin{bmatrix} u \\ v \\ \theta_A \\ \theta_B \end{bmatrix} = \begin{bmatrix} -l & -m & 0 & \dots & l & m & 0 \\ m & -l & 0 & \dots & -m & l & 0 \\ 0 & 0 & 1 & \dots & 0 & 0 & 0 \\ 0 & 0 & 0 & \dots & 0 & 0 & 1 \end{bmatrix} \begin{bmatrix} x_i \\ y_i \\ \theta_i \\ . \\ . \\ . \\ x_j \\ y_i \\ \theta_j \end{bmatrix}
$$

(10.33)

The member forces are the axial force, shear force, and end moments, which are vectorially equivalent to the member displacements, as shown. The elasticity equation relating axial force and extension has already been established. The other equations can be derived from the *slope-deflection* equations (4.3) and (4.4), as follows.

$$m_A = k(-3v/L + 2\theta_A + \theta_B)$$ (10.34)
$$m_B = k(-3v/L = \theta_A + 2\theta_B)$$ (10.35)

where $k = 2EI/L$. Taking moments about one end and substituting for m_A and m_B

$$q = k(6v/L^2 - 3\theta_A/L - 3\theta_B/L)$$ (10.36)

Hence the contribution of the member to the general elasticity equation $\mathbf{w} = \mathbf{ky}$ becomes

$$
\begin{bmatrix} p \\ q \\ m_A \\ m_B \end{bmatrix} = \begin{bmatrix} EA/L & 0 & 0 & 0 \\ 0 & 6k/L^2 & -3k/L & -3k/L \\ 0 & -3k/L & 2k & k \\ 0 & -3k/L & k & 2k \end{bmatrix} \begin{bmatrix} u \\ v \\ \theta_A \\ \theta_B \end{bmatrix}
$$

(10.37)

Matrix $\mathbf{k}$ for the whole structure has the same form as in equation (10.28) except that the single-element stiffnesses are replaced by sub-matrices.

Worked examples

10.6 Built-in beam

The built-in beam of Fig. 10.9 consists of two members of rigidity $6E3 \text{ kNm}^2$ and $12E3 \text{ kNm}^2$ welded together at B, where a couple of 100 kNm is applied. Use the matrix stiffness method to determine the deflection and rotation at B. Sketch the shear force and bending moment diagrams for the beam.

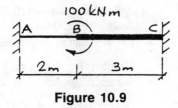

Figure 10.9

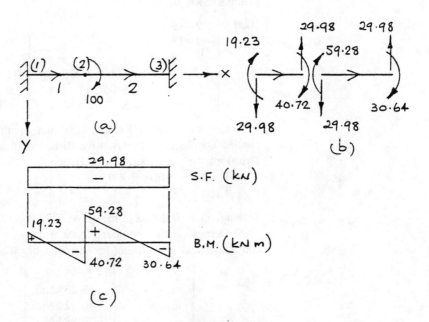

Figure 10.10

Solution — Fig. 10.10 (1) Number the joints and members and choose directions of orientation and of the general axes X, Y — Fig. 10.10(a).

(2) Construct $\mathbf{y} = \mathbf{AY}$. At joints (1) and (3) all displacements are restrained. At joint (2), if axial deformations are ignored, the only joint displacements are y_2 and θ_2; and the extension u can be omitted from the member displacement vector. Hence, using equation (10.33)

$$\begin{bmatrix} v_1 \\ \theta_{A1} \\ \theta_{B1} \\ v_2 \\ \theta_{A2} \\ \theta_{B2} \end{bmatrix} = \begin{bmatrix} 1 & 0 \\ 0 & 0 \\ 0 & 1 \\ -1 & 0 \\ 0 & 1 \\ 0 & 0 \end{bmatrix} \begin{bmatrix} y_2 \\ \theta_2 \end{bmatrix}$$

Note that for member 1 joint 2 is joint j; so that the elements of **A** come from columns 5 and 6 in equation (10.33). For member 2 joint 2 is joint i; so that the elements come from columns 2 and 3.

(3) Use equation (10.37) to construct matrix **k**, omitting the first row which deals with axial force and displacement. Then compute **kA** to provide the basis for subsequent determination of member forces.

Member 1: $L = 2$ m, $k = 2EI/L = 6000$ kNm
Member 2: $L = 3$ m, $k = 8000$ kNm

Hence, taking out 1000 as a common factor and noting that $l = 1$ for both members, **kA** becomes

$$1000 \begin{bmatrix} 9 & -9 & -9 & & & \\ -9 & 12 & 6 & & & \\ -9 & 6 & 12 & & & \\ & & & 16/3 & -8 & -8 \\ & & & -8 & 16 & 8 \\ & & & -8 & 8 & 16 \end{bmatrix} \begin{bmatrix} 1 & 0 \\ 0 & 0 \\ 0 & 1 \\ -1 & 0 \\ 0 & 1 \\ 0 & 0 \end{bmatrix} = \begin{bmatrix} 9 & -9 \\ -9 & 6 \\ -9 & 12 \\ -16/3 & -8 \\ 8 & 16 \\ 8 & 8 \end{bmatrix}$$

(4) Compute $\mathbf{K} = \mathbf{A'kA}$ and construct equation $\mathbf{W} = \mathbf{KY}$. Note that the applied couple is in the positive direction in accordance with the chosen directions for the general axes. Hence

$$\begin{bmatrix} 0 \\ 100 \end{bmatrix} = 1000 \begin{bmatrix} 43/3 & -1 \\ -1 & 28 \end{bmatrix} \begin{bmatrix} y_2 \\ \theta_2 \end{bmatrix}$$

Solving, $y_2 = 0.2498$E-3 m, $\theta_2 = 3.580$E-3 radians.

(5) Determine the member forces from $\mathbf{w} = \mathbf{kAY}$, using **kA** previously computed, and substituting for the joint displacements. Hence

$$\mathbf{w} = \begin{bmatrix} q_1 \\ m_{A1} \\ m_{B1} \\ q_2 \\ m_{A2} \\ m_{B2} \end{bmatrix} = \begin{bmatrix} -29.98 \text{ kN} \\ 19.23 \text{ kNm} \\ 40.72 \text{ kNm} \\ -29.98 \text{ kN} \\ 59.28 \text{ kNm} \\ 30.64 \text{ kNm} \end{bmatrix}$$

The directions of the member forces are as in Fig. 10.10(b).

(6) Sketch the shear force and bending moment diagrams − Fig. 10.10(c).

10.7 Distributed load

A symmetrical rectangular portal frame with fixed bases has height 6 m and span 12 m, and carries a uniformly distributed load of 20 kN/m on the beam. Given that all members have the same flexural rigidity EI, determine the principal displacements and bending moments.

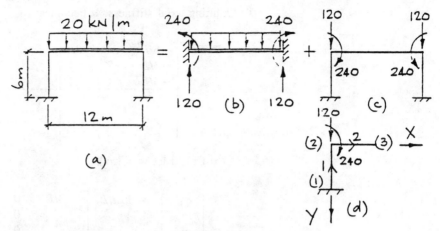

Figure 10.11

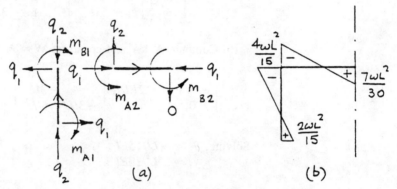

Figure 10.12

Solution — Figs 10.11 and 10.12   (1) Simulate the actual load-case Fig. 10.11(a) by superimposing the two cases (b) and (c), where (b) is the standard fixed end case. In (c) the moments and shears from (b) are reversed. Case (c), which no longer has a distributed load, is analysed by the matrix stiffness method.

(2) Since the frame is symmetrical it is only necessary to analyse the half frame in Fig. 10.11(d). All the displacements at joint (1) are restrained and, ignoring axial displacements and considering symmetry, $y_2=0$, $x_2=x_3=0$, $\theta_3=0$. Hence $\mathbf{Y} = \{\theta_2, y_3\}$.

(3) Write $\mathbf{y} = \mathbf{AY}$ by inspection of the figure,

$$
\begin{bmatrix}
v_1 \\
\theta_{A1} \\
\theta_{B1} \\
v_2 \\
\theta_{A2} \\
\theta_{B2}
\end{bmatrix}
=
\begin{bmatrix}
0 & 0 \\
0 & 0 \\
1 & 0 \\
0 & 1 \\
1 & 0 \\
0 & 0
\end{bmatrix}
\begin{bmatrix}
\theta_2 \\
y_3
\end{bmatrix}
$$

(4) Omitting axial stiffness, **k** becomes

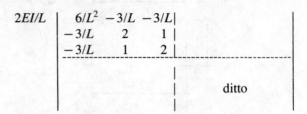

$$2EI/L \quad \begin{array}{|ccc}
6/L^2 & -3/L & -3/L \,| \\
-3/L & 2 & 1 \,| \\
-3/L & 1 & 2 \,| \\
\hline
& & | \\
& & | \qquad \text{ditto} \\
& & |
\end{array}$$

(5) Construct $\mathbf{w} = \mathbf{kAY}$

$$\begin{bmatrix} q_1 \\ m_{A1} \\ m_{B1} \\ q_2 \\ m_{A2} \\ m_{B2} \end{bmatrix} = 2EI/L \begin{bmatrix} -3/L & 0 \\ 1 & 0 \\ 2 & 0 \\ -3/L & 6/L^2 \\ 2 & -3/L \\ 1 & -3/L \end{bmatrix} \begin{bmatrix} \theta_2 \\ y_3 \end{bmatrix}$$

(6) Compute $\mathbf{A'kA}$ and construct $\mathbf{W} = \mathbf{A'kAY}$.

$$\begin{bmatrix} wL^2/3 \\ 0 \end{bmatrix} = 2EI/L \begin{bmatrix} 4 & -3/L \\ -3/L & 6/L^2 \end{bmatrix} \begin{bmatrix} \theta_2 \\ y_3 \end{bmatrix}$$

Solving, $\theta_2 = wL^3/15EI$
$\quad\quad\quad y_3 = wL^4/30EI$

(7) Back-substitute into $\mathbf{w} = \mathbf{kAY}$ for the member forces and add the standard results from the fixed-end case.

$$\begin{bmatrix} q_1 \\ m_{A1} \\ m_{B1} \\ q_2 \\ m_{A2} \\ m_{B2} \end{bmatrix} = wL^2/15 \begin{bmatrix} -6/L \\ 2 \\ 4 \\ 0 \\ 1 \\ -1 \end{bmatrix} + wL^2/6 \begin{bmatrix} 0 \\ 0 \\ 0 \\ 6/L \\ -2 \\ -1 \end{bmatrix} = \begin{bmatrix} -2wL/5 \\ 2wL^2/15 \\ 4wL^2/15 \\ wL \\ -4wL^2/15 \\ -7wL^2/30 \end{bmatrix}$$

(8) Determine the bending moments. The directions of the computed results are as shown in Fig. 10.12(a). As axial deformations were ignored in the analysis, the axial forces have to be determined by considering the equilibrium at joint 2. Thus the axial force in the beam must be equal and opposite to the shear force in the column and vice versa, as shown. The shear force q_2 in the beam refers to end A. At end B, i.e. mid-span in the beam, it decays to zero because of the uniformly distributed load. Using the convention positive curvature as concave, the bending moments are as shown in Fig. 10.12(b).

Stiffness matrix by direct construction

Since matrices **A** and **k** are large and sparse, arithmetic operations on them are impracticable by hand, even for the most trivial of structures. If performed by computer they would be grossly inefficient both in running time and storage. A better method is to operate on the contributions to these matrices made by each individual member. Thus by combining equations (10.33) and (10.37)

$$
\begin{bmatrix} p \\ q \\ m_A \\ m_B \end{bmatrix} = \begin{bmatrix} EA/L & 0 & 0 & 0 \\ 0 & 6k/L^2 & -3k/L & -3k/L \\ 0 & -3k/L & 2k & k \\ 0 & -3k/L & k & 2k \end{bmatrix} \begin{bmatrix} -l & -m & 0 & \dots & l & m & 0 \\ m & -l & 0 & \dots & -m & l & 0 \\ 0 & 0 & 1 & \dots & 0 & 0 & 0 \\ 0 & 0 & 0 & \dots & 0 & 0 & 1 \end{bmatrix} \begin{bmatrix} x_i \\ y_i \\ \theta_i \\ \cdot \\ \cdot \\ \cdot \\ x_j \\ y_j \\ \theta_j \end{bmatrix}
$$

$$(10.38)$$

Multiplying out

$$
\begin{aligned}
p &= EA/L \times [l(x_j-x_i) + m(y_j-y_i)] \\
q &= k[-6m(x_j-x_i)/L^2 + 6l(y_j-y_i)/L^2 - 3(\theta_i+\theta_j)/L] \\
m_A &= k[3m(x_j-x_i)/L - 3l(y_j-y_i)/L + 2\theta_i + \theta_j] \\
m_B &= k[3m(x_j-x_i)/L - 3l(y_j-y_i)/L + \theta_i + 2\theta_j]
\end{aligned}
$$

$$(10.39)$$

from which the forces in the member can be calculated when the joint displacements at its ends have been found. It is simpler, however, to determine the shear force directly by statics,

i.e. $\quad q = -(m_A + m_B)/L$ $\qquad\qquad$ (10.40)

Similarly, multiplication of **A′kA** for the member produces the following set of four sub-matrices which are its contribution to the stiffness matrix of the complete structure.

$$
\begin{bmatrix} \mathbf{K_{ii}} | \mathbf{K_{ij}} \\ \hline \mathbf{K_{ji}} | \mathbf{K_{jj}} \end{bmatrix} = \left[\begin{array}{ccc|ccc} a & c & d & -a & -c & d \\ c & b & e & -c & -b & e \\ d & e & f & -d & -e & g \\ \hline -a & -c & -d & a & c & -d \\ -c & -b & -e & c & b & -e \\ d & e & g & -d & -e & f \end{array} \right]
$$

$$(10.41)$$

where $\quad a = l^2 EA/L + 6m^2 k/L^2 \qquad b = m^2 EA/L + 6l^2 k/L^2$
$\qquad\quad c = lmEA/L - 6lmk/L^2 \qquad d = -3mk/L$
$\qquad\quad e = 3lk/L \qquad f = 2k \qquad\quad g = k$

The location of the sub-matrices within the equations **W** = **KY** when $i < j$ is as follows.

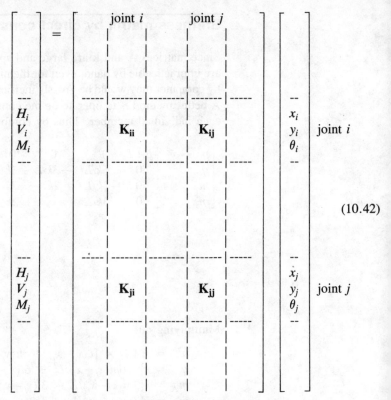

$$
\begin{array}{c}
\begin{array}{cc} \text{joint } i & \text{joint } j \end{array}
\end{array}
$$

$$
\begin{bmatrix} \\ \\ --- \\ H_i \\ V_i \\ M_i \\ --- \\ \\ \\ \\ \\ --- \\ H_j \\ V_j \\ M_j \\ --- \\ \\ \end{bmatrix}
=
\begin{bmatrix} & & \\ \mathbf{K_{ii}} & & \mathbf{K_{ij}} \\ & & \\ \mathbf{K_{ji}} & & \mathbf{K_{jj}} \\ & & \end{bmatrix}
\begin{bmatrix} \\ -- \\ x_i \\ y_i \\ \theta_i \\ -- \\ \\ \\ \\ \\ -- \\ \dot{x}_j \\ y_j \\ \theta_j \\ -- \\ \end{bmatrix}
\begin{array}{l} \\ \\ \\ \text{joint } i \\ \\ \\ \\ \\ \\ \\ \text{joint } j \\ \\ \end{array}
\tag{10.42}
$$

When $i > j$ the sub-matrices are exchanged diagonally.

Worked examples

10.8 Direct construction of the stiffness matrix

Given the numbering and orientations shown derive a set of equations from which the joint displacements of the structure in Fig. 10.13 may be determined. The members are identical, with $EA = 1.2E6$ kN and $EI = 19E6$ kNm^2.

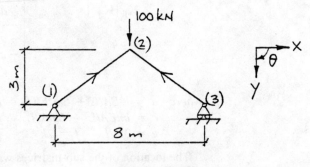

Figure 10.13

Solution (1) Locate within the stiffness matrix the sub-matrices contributed by each member. Use equation (10.42), noting that in the case of member 2 $i>j$, so the locations are exchanged diagonally.

$$
\begin{array}{l}
\text{joint 1} \\[4pt]
\text{joint 2} \\[4pt]
\text{joint 3}
\end{array}
\left|
\begin{array}{c|c|c}
(\mathbf{K_{ii}})_1 & (\mathbf{K_{ij}})_1 & \\
\hline
(\mathbf{K_{ji}})_1 & (\mathbf{K_{jj}})_1+(\mathbf{K_{ii}})_2 & (\mathbf{K_{ji}})_2 \\
\hline
 & (\mathbf{K_{ij}})_2 & (\mathbf{K_{ii}})_2
\end{array}
\right|
$$

(2) Construct $\mathbf{W} = \mathbf{KY}$ by filling in the elements of the sub-matrices from equation (10.41). Taking the restraints at joints 1 and 3 into account, $Y = \{\theta_1, x_2, y_2, \theta_2, x_3, \theta_3\}$. Rows and columns in the sub-matrices corresponding to restrained displacements are omitted. Hence

$$
\begin{bmatrix} 0 \\ \hline 0 \\ 100 \\ 0 \\ \hline 0 \\ 0 \end{bmatrix}
=
\left[
\begin{array}{c|cccc|cc}
f_1 & -d_1 & -e_1 & g_1 & & \\
\hline
-d_1 & a_1+a_2 & c_1+c_2 & -d_1-d_2 & -a_2 & -d_2 \\
-e_1 & c_1+c_2 & b_1+b_2 & -e_1-e_2 & -c_2 & -e_2 \\
g_1 & -d_1-d_2 & -e_1-e_2 & f_1+f_2 & d_2 & g_2 \\
\hline
 & -a_2 & -c_2 & d_2 & a_2 & d_2 \\
 & -d_2 & -e_2 & g_2 & d_2 & f_2
\end{array}
\right]
\begin{bmatrix} \theta_1 \\ \hline x_2 \\ y_2 \\ \theta_2 \\ \hline x_3 \\ \theta_3 \end{bmatrix}
$$

(3) Calculate the values of the elements. Working in kN,m units,
Member 1 $l = 0.8, m = -0.6, L = 5$. Hence $k = 7.6\text{E}3$, $EA/L = 240\text{E}3$
Member 2 $l = -0.8$, otherwise as member 1.
Hence:

$$
\begin{aligned}
a_1 &= a_2 = 154.3\text{E}3 & d_1 &= d_2 = 2.736\text{E}3 & f_1 &= f_2 = 15.20\text{E}3 \\
b_1 &= b_2 = 87.57\text{E}3 & e_1 &= -e_2 = 3.648\text{E}3 & g_1 &= g_2 = 7.600\text{E}3 \\
c_2 &= -c_1 = 114.3\text{E}3
\end{aligned}
$$

10.9 Equal displacements

(a) Show that the displacement vector for the portal frame in Fig. 10.14 can be reduced to $\{x_2, \theta_2\}$ if axial deformations are neglected.

(b) Given that Young's modulus is 200 kN/mm^2 and second moments of area are 6E6 mm^4 (columns) and 30E6 mm^4 (beam), determine the joint displacements and member forces, and sketch the bending moment diagram of the frame.

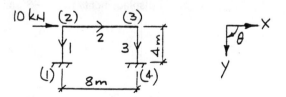

Figure 10.14

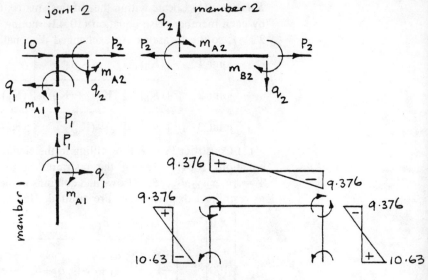

Figure 10.15

Solution (a) — Fig. 10.14 If axial deformations are neglected, then, as there is no change in the length of the columns or the beam, $y_2 = y_3 = 0$ and $x_2 = x_3$. Furthermore, since the columns are identical, $\theta_2 = \theta_3$. Therefore the joint displacements are fully defined by $\mathbf{Y} = \{x_2, \theta_2\}$.

Solution (b) (1) Construct $\mathbf{W} = \mathbf{KY}$. If equal displacements are included initially,

$$
\begin{bmatrix} \mathbf{W_2} \\ \hline \mathbf{W_3} \end{bmatrix} = \begin{bmatrix} (\mathbf{K_{ii}})_1 + (\mathbf{K_{ii}})_2 & (\mathbf{K_{ij}})_2 \\ \hline (\mathbf{K_{ji}})_2 & (\mathbf{K_{jj}})_2 + (\mathbf{K_{ii}})_3 \end{bmatrix} \begin{bmatrix} \mathbf{Y_2} \\ \hline \mathbf{Y_3} \end{bmatrix}
$$

where $\mathbf{Y_2} = \{x_2, \theta_2\}$ and $\mathbf{Y_3} = \{x_3, \theta_3\}$; $\mathbf{W_2}$ and $\mathbf{W_3}$ contain the corresponding loads. i.e. from equation (10.41)

$$
\begin{bmatrix} 10 \\ 0 \\ 0 \\ 0 \end{bmatrix} = \begin{bmatrix} a_1 + a_2 & d_1 + d_2 & -a_2 & d_2 \\ d_1 + d_2 & f_1 + f_2 & -d_2 & g_2 \\ -a_2 & -d_2 & a_3 + a_2 & d_3 - d_2 \\ d_2 & g_2 & d_3 - d_2 & f_3 + f_2 \end{bmatrix} \begin{bmatrix} x_2 \\ \theta_2 \\ x_3 \\ \theta_3 \end{bmatrix}
$$

(2) Reduce the stiffness matrix, first by adding the rows corresponding to equal displacement, i.e. $1+3$ and $2+4$

$$
\begin{bmatrix} 10+0 \\ 0+0 \end{bmatrix} = \begin{bmatrix} a_1 & d_1 & a_3 & d_3 \\ d_1 + 2d_2 & f_1 + f_2 + g_2 & d_3 - 2d_2 & g_2 + f_3 + f_2 \end{bmatrix} \begin{bmatrix} x_2 \\ \theta_2 \\ x_3 \\ \theta_3 \end{bmatrix}
$$

Finally, the effect of replacing x_3 by x_2 and θ_3 by θ_2 is obtained by adding columns 1 to 3 and 2 to 4, thus

$$\begin{bmatrix} 10 \\ 0 \end{bmatrix} = \begin{bmatrix} a_1+a_3 & d_1+d_3 \\ d_1+d_3 & f_1+2f_2+2g_2+f_3 \end{bmatrix} \begin{bmatrix} x_2 \\ \theta_2 \end{bmatrix}$$

(3) Compute the elements, working in kN, mm units; and solve the equations.
Members 1 and 3: $l = 0$, $m = 1$, $k = 0.6E6$
Member 2: $l = 1$, $m = 0$, $k = 1.5E6$
Ignoring axial stiffnesses,

$$a_1+a_3 = 0.450 \qquad d_1+d_3 = -900$$
$$f_1+f_3 = 2.4E6 \qquad 2f_2+2g_2 = 9.6E6$$

Hence .

$$\begin{bmatrix} 10 \\ 0 \end{bmatrix} = \begin{bmatrix} 450E\text{-}3 & -900 \\ -900 & 11.4E6 \end{bmatrix} \begin{bmatrix} x_2 \\ \theta_2 \end{bmatrix}$$

Solving, $x_2 = 26.39$ mm. $\theta_2 = 2.089E\text{-}3$ rad
(4) Determine the end moments for member 1 from equation (10.39).

$x_j - x_i = -x_2 = -26.39$, $\theta_i = \theta_2 = 2.083E\text{-}3$, other terms are zero.
Hence

$$m_{A1} = k[3m(x_j-x_i)/L + 2\theta_i] = -9.375E3 \text{ kNmm}$$
$$m_{B1} = k(3m(x_j-x_i)/L + \theta_i] = -10.625E3 \text{ kNmm}$$

The other member forces can be derived from skew-symmetry and by considering the equilibrium of joint 2 and member 2 — Fig. 10.15. All the member forces are shown in the positive direction.

$$q_1 = q_3 = 10/2 = 5 \text{ kN}$$
$$m_{A2} = m_{B2} = -m_{A1} = 9.376E3 \text{ kNmm}$$
$$q_2 = -(m_{A2}+m_{B2})/L = -2\times9.376E3/8E3 = -2.344 \text{ kN}$$
$$p_1 = -q_2 = 2.344 \text{ kN}$$
$$p_2 = q_1-10 = -5 \text{ kN}$$

(5) Sketch the bending moment diagram — Fig. 10.15.

10.10 Linearly related displacements

(a) Prove that, if axial deformations are neglected, the vertical deflection at the apex of the pitched-roof portal frame in Fig. 10.16 is twice the horizontal deflection at the eaves.
(b) Show that the vector $Y = \{x_2,\theta_2\}$ fully defines the displacements of the frame.
(c) Given that the flexural rigidities are 120E3 kNm² (rafters) and 60E3 kNm² (columns), determine the joint displacements.

Solution (a)   The apex and eaves joints are connected by members 2 and 3. Take member 2 and set $u = 0$ in equation (10.33), noting that $i = 3$ and $j = 2$,

$$0 = -l_2x_3 - m_2y_3 + l_2x_2 + m_2y_2$$

But, from symmetry $x_3=0$, and if there is to be no axial deformation of the column, $y_2=0$. Hence

$$y_3 = x_2l_2/m_2 = x_2\cot\alpha_2 = 2x_2 \dots \text{Q.E.D.}$$

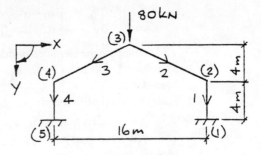

Figure 10.16

Solution (b) From symmetry $x_3 = \theta_3 = 0$, $x_4 = -x_2$, and $\theta_4 = -\theta_2$. Therefore, since $y_3 = 2x_2$ and all other displacements are zero, the displacements are fully defined by $Y = \{x_2, \theta_2\}$.

Solution (c) (1) Construct the equations $W = KY$ directly for the right hand half of the frame, divided at the apex and carrying half the vertical load. Include joint 3 at this stage.

$$\begin{bmatrix} W_2 \\ \hline W_3 \end{bmatrix} = \begin{bmatrix} (K_{ii})_1 + (K_{jj})_2 & | & (K_{ji})_2 \\ \hline (K_{ij})_2 & | & (K_{ii})_2 \end{bmatrix} \begin{bmatrix} Y_2 \\ \hline Y_3 \end{bmatrix}$$

Assuming initially the joint displacement vector $\{x_2, \theta_2, y_3\}$, then from equation (10.41)

$$\begin{bmatrix} 0 \\ 0 \\ \hline 40 \end{bmatrix} = \begin{bmatrix} a_1 + a_2 & d_1 - d_2 & | & -c_3 \\ d_1 - d_2 & f_1 + f_2 & | & e_2 \\ \hline -c_2 & e_2 & | & b_2 \end{bmatrix} \begin{bmatrix} x_2 \\ \theta_2 \\ \hline y_3 \end{bmatrix}$$

The solution of these equations as they stand requires that the axial deformation of the rafter be included. Otherwise the stiffness matrix must be reduced by means of the relationship developed in part (a). First, since y_3 is to be replaced by x_2, it is necessary to determine a hypothetical load H in the direction of x_2 that will do the same work as the actual 40 kN in the direction of y_3. Hence

$$40y_3 = Hx_2$$

replacing y_3 by $2x_2$

$$H = 80 \text{ kN}$$

Proceeding as in the previous problem the bottom row must be multiplied by 2 and added to the first row, thus

$$\begin{bmatrix} 0 + 80 \\ 0 \end{bmatrix} = \begin{bmatrix} a_1 + a_2 - 2c_2 & d_1 - d_2 + 2e_2 & -c_2 + 2b_2 \\ d_1 - d_2 & f_1 + f_2 & e_2 \end{bmatrix} \begin{bmatrix} x_2 \\ \theta_2 \\ y_3 \end{bmatrix}$$

Now, since $y_3 = 2x_2$, the elimination of y_3 from the displacement vector is achieved by multiplying column 3 by 2 and adding to column 1, thus

$$\begin{bmatrix} 80 \\ 0 \end{bmatrix} = \begin{bmatrix} a_1+a_2-4c_2+4b_2 & d_1-d_2+2e_2 \\ d_1-d_2+2e_2 & f_1+f_2 \end{bmatrix} \begin{bmatrix} x_2 \\ \theta_2 \end{bmatrix}$$

(2) Compute the elements from equation (10.41) in Kn,m units, ignoring the terms involving EA/L, and solve the equations.

Member 1: $L=4$, $l=0$, $m=1$, $k=30E3$
Member 2: $L=\sqrt{(8^2+4^2)}=8.944$, $l=8/L=0.8944$, $m=4/L=0.4472$,
$\qquad\quad k=26.83E3$

$$\begin{array}{lll}
a_1 = 11.25E3 & a_2 = 0.4025E3 & d_2 = -4.025E3 \\
d_1 = -22.5E3 & b_2 = 1.610E3 & e_2 = 8.049E3 \\
f_1 = 60.00E3 & c_2 = -0.8049E3 & f_2 = 53.66E3
\end{array}$$

Hence

$$\begin{bmatrix} 80 \\ 0 \end{bmatrix} = \begin{bmatrix} 21.31E3 & -2.377E3 \\ -2.377E3 & 113.7E3 \end{bmatrix} \begin{bmatrix} x_2 \\ \theta_2 \end{bmatrix}$$

Solving, $x_2 = 3.762E-3$ m, $\theta_2 = 78.7E-6$ rad, $y_3 = 2x_2 = 7.524E-3$ m

---

Computer programs

The following programs are written in *Mallard-80* BASIC and have been tested on an *Amstrad PCW 8256* computer with a free memory of 31 597 bytes. Except where specifically mentioned extended facilities of this version of BASIC have been avoided, so the listings should be compatible with any BASIC which allows variable names consisting of two significant characters.

The instructions for the input of data appear on the screen and are self explanatory. Input instructions conclude with a ? indicating that an item of data should be entered, i.e. the item should be typed in, followed by pressing the 'ENTER' or 'RETURN' key as appropriate. No instructions as to units are given but any consistent set can be used. All output is to the screen and breaks are included so that information does not run off the screen. Examples of typical runs are given.

Flexibility method

There are two programs: for the analysis of pin-jointed and rigidly jointed structures. Combinations of pinned and rigid joints in the same structure are not allowed. The programs are intended chiefly for the checking of problems that have been worked by hand. The number of members is limited to 20 and the number of redundant restraints to 10. These numbers can be changed in the dimension and input statements to suit individual requirements. However, readers who require to analyse larger structures for a specific purpose, such as structural design, are advised to use the program based on the stiffness method.

After the input of general data the member flexibilities L/EA for pinned joints and $L/6EI$ for rigid joints are input, followed by the equilibrium equations $\mathbf{w} = \mathbf{BW}$ which must be constructed by hand. Matrix $\mathbf{B}$ is input column by column. Members must be numbered sequentially and the rows of $\mathbf{B}$ must be in the same sequence. In the case of rigid joints there are two rows of $\mathbf{B}$ for each member, corresponding to the ends A and B. The flexibility matrix $\mathbf{F} = \mathbf{B}'\mathbf{fB}$ is then computed, followed by instructions for the input of the external loads and any lack-of-fit or settlement displacements. The load vector $\mathbf{W}$ is organised so that the redundant restraints are the last items, as in the hand method previously described.

First output is the flexibility matrix $\mathbf{F}$, column by column. This section of the program is written as a subroutine, starting at line 1000 and can be omitted, together with the GOSUB instruction in line 610, if not required. The remainder of the output is as follows.

(1) Redundant restraints in the order in which the corresponding columns of $\mathbf{B}$ were input.

(2) Member forces.

(3) Displacements in the same order and direction as the external loads in $\mathbf{W}$.

Programming is simplified in the case of rigidly jointed members by dividing matrix $\mathbf{B}$ into two arrays BA and BB which contain the influence coefficients corresponding to ends A and B of the members. The equations in the restraints are solved by a simple full-matrix subroutine using Gaussian elimination.

Stiffness method

This program is capable of analysing both rigidly jointed and pin-jointed structures up to a maximum size of 20 joints and 30 members. Larger structures could probably be accommodated within the available memory, however, and readers are invited to experiment by altering the dimension statements. More information on array storage is given later.

N.B. The statements in lines 20 and 30 may not be available in all versions of BASIC. Line 20 sets the lowest array subscript to unity instead of the more usual zero. This is to avoid wasting memory and can be omitted without affecting the running of the program. Line 30 defines as type integer any variables with names beginning with the letters H to N inclusive, S to V inclusive, and Z. This again is to save memory, and to promote faster running, but can be omitted. Alternatively the variables can be defined individually as type integer by appending a % symbol to each variable name.

Rules for the preparation of data are as follows.

(1) Draw a line diagram of the structure and choose general axes. It is preferable, but not essential, to place the origin of the axes so that the coordinates of all the joints are positive. The axes may have any orientation provided that they are a right handed set, i.e. the positive direction of θ is from X to Y.

(2) Put in joints at changes of direction or section and at load points, and number them sequentially, starting from 1. Running time and memory requirements are reduced by keeping the difference between the numbers allocated to the joints at the ends of each member small.

(3) Number members sequentially, starting from 1, and mark the directions of orientation of their axes. Allocation of numbers and orientations is arbitrary and has no effect on the running of the program.

(4) Arrange data in the following groups. Any units can be used provided that they are consistent. This avoids the need to enter the units as data.

General data: Young's modulus; number of members; number of joints; number of loaded joints.

Joint data: For each joint — the coordinates (X,Y); the degrees of freedom (D.O.F.) in the X,Y, and θ directions (1 = freedom to move, 0 = restraint).

Member data: Cross sectional area (C.S.A.); second moment of area (S.M.A.); joint *i* (end A); joint *j* (end B), as indicated by the orientation arrows.

Load data: Loaded joint number; X load; Y load; moment load. Loads are positive in the directions of X, Y and θ defined by the general axes.

The program breaks into direct mode after the entry of joint, member and load data to allow for correction, if necessary — see Solution 10.12.

Output is to the screen and is self explanatory. Sign convention for joint displacements is as defined by the general axes. Member end moments M_A and M_B are positive in the direction of θ, axial force P is positive when tensile, and the positive direction of the shear force Q is obtained by rotating through 90° from P in the direction of θ. Two examples with different orientations of the general axes are given in Fig. 10.17.

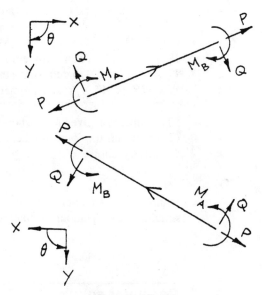

Figure 10.17

Programming notes   The analysis is by direct construction of the stiffness matrix, as described earlier. Programming is complicated by the necessity of dealing with joints which have a variety of restraints and by the compact method adopted for the storage and solution of the equations $\mathbf{W} = \mathbf{KY}$. Since the stiffness matrix is symmetrical only the elements in each row up to and including the element on the leading diagonal are stored, starting from the first essentially non-zero element. The stiffness matrix is stored sequentially in the one-dimensional array A and, since the number of elements in each row can vary, the addresses of the leading-diagonal elements in the sequence are stored in another one-dimensional array S. This method of solution, which was devised by Jennings (1966), reduces the running time and storage as compared with full matrix techniques, the advantages increasing with the number of equations. It also has the advantage over a simple band-matrix solution that the storage is not greatly increased by an unavoidably large difference between the joint numbers at the ends of a single member.

All the arrays are one-dimensional. Their description and storage requirements, assuming N joints and M members, are as follows.

Array	Description of array (or kth element)	Dimension
NX,NY,NT	Degrees of freedom (X,Y,θ) of joint k	N
X,Y	Coordinates (X,Y) of joint k	N
LJ	Lowest joint number connected to joint k	N
NC	Number of rows in **K** before joint k	N
A	Compact, sequential form of **K**	42N*
S	Addresses of leading diagonal elements in A	3N
B	Load vector (and joint displacements)	3N
C9	Working store in the solution subroutine	3N
CA,CI	C.S.A. and S.M.A. of member k	M
JI,JJ	Joint numbers i and j for member k	M
J	Loaded joint number (kth line of load data)	10†
WX,WY,WM	X,Y and moment loads at joint J(k)	10†

*The dimension given for A above is an estimate based on an average difference of 4 between the joint numbers at member ends.
†Arbitrary choice.

Reference

Jennings, A. A compact storage scheme for the solution of symmetric linear simultaneous equations. *Computer Journal*, 1966, **9**, November.

Computer solutions – flexibility method

10.11 Pin-jointed frame

Determine the forces in the members, the vertical deflection at D and the horizontal movement at C in the pin-jointed frame of Fig. 10.18(a) given Young's modulus 200 kN/mm², cross sectional areas: horizontal members 400 mm², vertical and diagonal members 300 mm². Assume that there is no settlement, but that on assembly member BE was 2 mm too short.

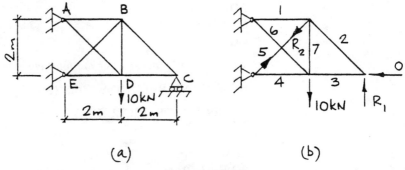

(a)   (b)

Figure 10.18

Solution (1) Number the members sequentially, in any order – Fig. 10.18(b).
(2) Select redundancies, say R_1 (reaction at C) and R_2 (force in member 5).
(3) Determine the flexibilities L/EA. Use kN,mm units.

Member	1	2	3	4	5	6	7
L/EA	0.025	0.04714	0.025	0.025	0.04714	0.04714	0.03333

(4) Construct the equations $\mathbf{w} = \mathbf{BW}$ by hand, including a dummy horizontal load at C for the horizontal deflection.

$$\begin{bmatrix} p_1 \\ p_2 \\ p_3 \\ p_4 \\ p_5 \\ p_6 \\ p_7 \end{bmatrix} = \begin{bmatrix} 0 & 0 & -1 & -0.7071 \\ 0 & 0 & -1.414 & 0 \\ 0 & -1 & 1 & 0 \\ -1 & -1 & 2 & -0.7071 \\ 0 & 0 & 0 & 1 \\ 1.414 & 0 & -1.414 & 1 \\ 0 & 0 & 1 & -0.7071 \end{bmatrix} \begin{bmatrix} 10 \\ 0 \\ R_1 \\ R_2 \end{bmatrix}$$

(5) Run the program MFLEX1 and enter the data in response to the screen prompts, as follows. (Screen prompts are given in upper case, data entries in bold type, comments not appearing on the screen in italics.)

HOW MANY MEMBERS (MAX 20)? **7**
ENTER MEMBER FLEXIBILITIES L/EA

MEMBER 1? **.025**
MEMBER 2? **.04714**
etc.

ENTER COLS OF MATRIX B
HOW MANY COLUMNS (MAX 20)? **4**

COLUMN 1
MEMBER 1? **0**
MEMBER 2? **0**
etc.

COLUMN 2
MEMBER 1? **0**
etc.

There will be a pause after the final entry while **B′fB** *is computed.*

HOW MANY REDUNDANCIES (MAX 10)? **2**

ENTER LOADS, INCLUDING ZEROS

ROW 1? **10**
ROW 2? **0**

The computer asks here for the known loads in the order in which they appear in the load vector, i.e. the actual load at D and the dummy at C.

ENTER LACK OF FIT AND SETTLEMENT

The computer is now asking for the displacements in rows 3 and 4 of the displacement vector, i.e the settlement at C and the lack of fit in member 5. Settlement is positive in the direction of R_1; *lack of fit is positive when the member is too short. Units must be consistent, i.e. mm in this case.*

ROW 3? **0**
ROW 4? **2**

End of data entry. The columns of the flexibility matrix are now output. After rounding the computer output the result is as follows.

$$\begin{bmatrix} 0.1193 & 0.025 & -0.1443 & 0.08433 \\ 0.0250 & 0.050 & -0.075 & 0.01768 \\ -0.1443 & -0.075 & 0.3718 & -0.1079 \\ 0.08433 & 0.01768 & -0.1079 & 0.1359 \end{bmatrix}$$

The redundant restraints are then output in the order in which they appear in the load vector.

REDUNDANT FORCES ARE:
 8.25
15.06

followed by

MEMBER FORCES ARE:
MEMBER 1   −18.99
MEMBER 2   −11.66
MEMBER 3   8.25
MEMBER 4   − 4.15
MEMBER 5   15.05
MEMBER 6   17.53
MEMBER 7   − 2.40

The forces are in kN. The computer output has been rounded to two decimal places.

DEFLECTIONS ARE:
   1.27
 −0.10
(in mm, rounded to two decimal places)

End of run.

10.12 Rigidly jointed frame

Check the solution of Example 10.4 by computer.

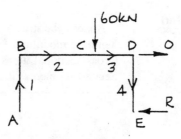

Figure 10.19

Solution − Fig. 10.19 (1) Number the members in the same order as in the problem.

(2) Run the program MFLEX2 and respond to the prompts as follows. (Screen prompts are given in upper case, data entries in bold type, comments not appearing on the screen in italics.)

HOW MANY MEMBERS (MAX 20)? **4**

ENTER MEMBER FLEXIBILITIES L/6EI *(From Example 10.4)*
MEMBER 1? **62.5E-6**
MEMBER 2   **40.0E-6**
MEMBER 3   **20.0E-6**
MEMBER 4   **62.5E-6**

ENTER COLS OF MATRIX B
HOW MANY COLUMNS (MAX 20)? **3**

COLUMN 1
MEMBER 1

END A? **0**
END B? **0**

MEMBER 2

END A?   **0**
END B? **−2**

etc.
   The remainder of the input is in the same form as in the previous problem.

Output also except that member forces are in two columns, giving the moments at ends A and B.

Computer solutions — stiffness method

Note: Since the computer analysis takes axial deformations into account, the solutions of rigidly jointed structures will differ slightly from the solutions obtained by hand methods.

10.13 Rigidly jointed frame

Analyse the structure in Fig. 10.20(a), given the following general data.
Young's modulus: 200 kN/mm^2
Cross sectional areas: AB 8540 mm^2, others $12\,930 \text{ mm}^2$
Second moments of area: AB $294.0\text{E}6 \text{ mm}^4$, others $616.6\text{E}6 \text{ mm}^4$

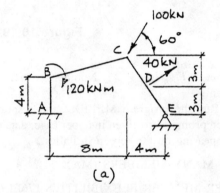

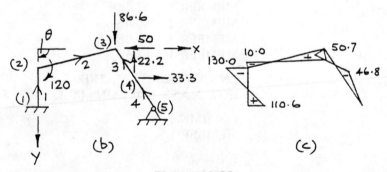

Figure 10.20

Solution — Fig. 10.20(b) and (c) (1) Allocate joint and member numbers and member orientations. The joint numbering shown gives the minimum joint number difference for all members.

(2) Choose general axes. With the origin as shown, all joint coordinates are positive.

(3) Resolve the loads into components parallel to the general axes.

(4) Select units. Since kNm is generally regarded as the most convenient unit for moments, use kN and m.

(5) Prepare a data sheet as follows. (Data to be entered is shown in bold type, the names of the arrays containing the data in brackets.) Then run the program MSTIFF and enter the data in response to the screen prompts.

General data

Young's modulus: **200E6** kN/m^2
Number of members: **4**
Number of joints: **5**
Number of loaded joints: **3**

Joint data

Joint No.	Coordinates		Degrees of freedom (D.O.F.)		
			X	Y	θ
	(X)	(Y)	(NX)	(NY)	(NT)
1	**0**	**6**	**0**	**0**	**0**
2	**0**	**2**	**1**	**1**	**1**
3	**8**	**0**	**1**	**1**	**1**
4	**10**	**3**	**1**	**1**	**1**
5	**12**	**6**	**0**	**0**	**1**

⟨*Break into direct mode*⟩
Member data

Member No.	C.S.A.	S.M.A.	Joint *i*	Joint *j*
	(CA)	(CI)	(JI)	(JJ)
1	**.00854**	**.000294**	**1**	**2**
2	**.01293**	**.0006166**	**2**	**3**
3	**.01293**	**.0006166**	**4**	**3**
4	**.01293**	**.0006166**	**5**	**4**

⟨*Break into direct mode*⟩
Load data (positive directions are X,Y,θ of the general axes)

Loaded joint No. (J)	X load (WX)	Y load (WY)	M load (WM)
2	**0**	**0**	**120**
3	**−50**	**86.6**	**0**
4	**33.3**	**−22.2**	**0**

⟨*Break into direct mode*⟩

When BASIC breaks into direct mode incorrectly entered items of data can be altered, using the array names in brackets. For example, suppose that the second moment of area for member 3 had been entered as .006166. Carry on entering data until the DIRECT MODE message appears on the screen, then type the statement

CI(3) = .0006166

followed by pressing the ENTER or RETURN key, as appropriate. When all corrections to the member data have been made, or if there are no corrections, return to the program by typing CONT, following by pressing ENTER or RETURN.

The procedure is slightly more complicated in the case of load data. For example, suppose the negative sign had been omitted from the X load at joint 3. After entering the rest of the load data, when the break into direct mode occurs, enter

WX(2) = −50

The array subscript is 2 because, although it refers to joint 3, it is in the second line of load data.

There are pauses in the running of the program after the entry of member data, while the stiffness matrix is being constructed and packed into array A, and after the entry of the load data, while the equations equivalent to $W = KY$ are being solved and member forces calculated. The results are then output, with breaks to allow copying from the screen. The following is a summary of the output.

Joint displacements (m,rad)

Joint	X	Y	θ
1	0.0	0.0	0.0
2	−4.135E-3	5.299E-5	6.608E-4
3	−3.693E-3	2.660E-3	−6.979E-4
4	−8.480E-4	6.640E-4	−7.542E-4
5	0.0	0.0	−6.970E-5

Member forces (kN,kNm)

Member	P	Q	M_A	M_B
1	−22.63	−60.16	110.6	130.0
2	−63.85	7.36	−10.03	−50.66
3	−58.86	−27.04	46.82	50.66
4	−58.86	−12.98	0.0	−46.82

Using the convention positive curvature is concave, i.e. bending is towards the inside of the frame, the bending moment diagram is as shown in Fig. 10.20(c).

Analyse the structure in Fig. 10.21(a), given Young's modulus 200E6 kN/m², cross sectional area 10E-3 m², second moment of area 500E-6 m⁴. Sketch the bending moment diagram.

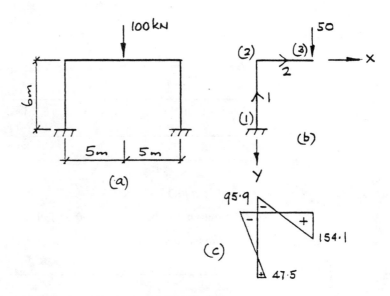

Figure 10.21

Solution — Fig. 10.21(b) and (c) (1) Since the structure is symmetrical analyse the half frame in Fig. 10.21(b), which carries half the central load.

(2) Prepare data as follows. The main point to note is the degree of freedom of joint 3. Because of symmetry, translation in the X direction and rotation are both zero.

General data
Young's modulus: **200E6** kN/m²
Number of members: **2**
Number of joints: **3**
Number of loaded joints: **1**

Joint data

Joint No.	Coordinates		Degrees of freedom (D.O.F.)		
			X	Y	θ
	(X)	(Y)	(NX)	(NY)	(NT)
1	0	6	0	0	0
2	0	0	1	1	1
3	5	0	0	1	0

Member data

Member No.	C.S.A. (CA)	S.M.A. (CI)	Joint i (JI)	Joint j (JJ)
1	10E-3	500E-6	1	2
2	10E-3	500E-6	2	3

Load data (positive directions are X, Y, θ of the general axes)

Loaded joint No. (J)	X load (WX)	Y load (WY)	M load (WM)
3	0	50	0

⟨*End of data*⟩

The output can be summarised as follows.

Joint displacements (m,rad)

Joint	X	Y	θ
1	0.0	0.0	0.0
2	5.97E-5	150.0E-6	1.45E-3
3	0.0	8.99E-3	0.0

Member forces (kN,kNm)

Member	P	Q	M_A	M_B
1	−50.00	−23.90	47.46	95.92
2	−23.90	50.00	−95.92	−154.08

From the general axes θ and moments are positive clockwise. The bending moment diagram is shown in Fig. 10.21(c).

10.14 Distributed load

Analyse the frame of the previous problem given that the point load is replaced by a total distributed load of 180 kN.

Solution − Fig. 10.22 (1) Determine the fixed-end moments and shears for the beam from standard results, i.e. $M = WL/12$, $Q = WL/2$, giving the

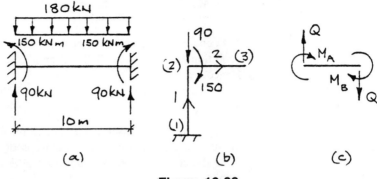

Figure 10.22

results in Fig. 10.22(a). Reverse and apply these to the half frame in Fig. 10.22(b).

(2) Prepare data. The only difference from the previous problem is in the load data, which is as follows.

Load data (positve directions are X,Y,θ of the general axes)

Loaded joint No. (J)	X load (WX)	Y load (WY)	M load (WM)
2	0	90	150

The output can be summarised as follows.

Joint displacements (m,rad)

Joint	X	Y	θ
1	0.0	0.0	0.0
2	7.17E-5	0.27E-3	1.74E-3
3	0.0	4.63E-3	0.0

Member forces (kN,kNm)

Joint	P	Q	M_A	M_B
1	−90.00	−28.70	56.96	115.10
2	−28.70	0.00	34.89	−34.89

(3) For member 2 add the results of the fixed-end condition to the computed results, noting that the positive directions of the computed results are as shown in Fig.10.22(c). Hence at end A

$$M_A = 34.89 - 150 = -115.1\,\text{kNm}$$
$$Q = 0 + 90 = 90\,\text{kN}$$

At end B (mid-span of the beam) the bending moment under the fixed-end condition is $WL/24$, i.e. 75 kNm anticlockwise.
Hence
$$M_B = -34.89 - 75 = -109.9\,\text{kNm}$$
and $Q_B = Q_A - \text{load to mid-span} = 90 - 90 = 0$

(4) Determine the displacements in a similar manner. At the ends of the beam the fixed-end condition, by definition, produces no displacements, but at joint 3 the deflection is $WL^3/384EI$, i.e. 4.69E-3 m.
Hence
$$Y_3 = 4.63\text{E-3} + 4.69\text{E-3} = 9.32\text{E-3 m}$$

(5) Applying the usual convention for bending moments the results are
$$M_1 = 56.96,\ M_2 = -115.1,\ M_3 = 109.9\ \text{kNm}$$

10.15 Pin-jointed frame

Analyse the frame in Example 6.1, using the matrix stiffness method.

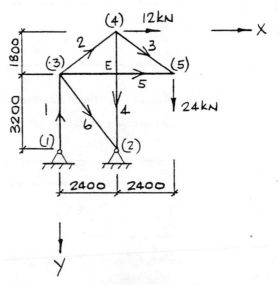

Figure 10.23

Solution — Fig. 10.23 Pin-jointed frames can be analysed with the program MSTIFF by setting the rotational degree of freedom of each joint and the second moment of area of each member to zero. In this problem there is no need to prepare data for joint E, at which members intersect at right angles, unless its displacement is required. The four members intersecting at E can therefore be treated as two.

(1) Prepare data in kN,mm units and run the program.

General data

Young's modulus: **200** kN/mm^2

Number of members: **6**

Number of joints: **5**

Number of loaded joints: **2**

Joint data

Joint No.	Coordinates		Degrees of freedom (D.O.F.)		
			X	Y	θ
	(X)	(Y)	(NX)	(NY)	(NT)
1	0	3200	0	0	0
2	2400	3200	0	0	0
3	0	0	1	1	0
4	2400	1800	1	1	0
5	4800	0	1	1	0

Member data

Member No.	C.S.A.	S.M.A.	Joint *i*	Joint *j*
	(CA)	(CI)	(JI)	(JJ)
1	1000	0	1	3
2	1000	0	3	4
3	1000	0	4	5
4	1000	0	4	2
5	1000	0	3	5
6	1000	0	3	2

Load data (positive directions are X,Y,θ of the general axes)

Loaded joint No.	X load	Y load	M load
(J)	(WX)	(WY)	(WM)
4	12	0	0
5	0	24	0

⟨*End of data*⟩

Results are:

Joint displacements (mm)

Joint	X	Y
1	0.0	0.0
2	0.0	0.0
3	1.712	−0.784
4	4.40	1.42
5	0.944	7.033

Member forces (kN)

Member	P
1	49.0
2	55.0
3	40.0
4	−57.0
5	−32.0
6	−20.0

Worked example

10.16 Joint numbering

Two sets of joint numbers are given for the structure in Fig. 10.24. Assuming the column bases to be (a) fixed, (b) pinned, which set would give the more compact band of elements in the stiffness matrix? Sketch the shape of the stiffness matrix in each case and determine the number of elements, assuming that in each row only the elements starting from the first essentially non-zero element up to and including the leading diagonal element are stored.

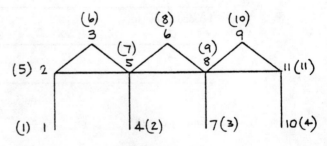

Figure 10.24

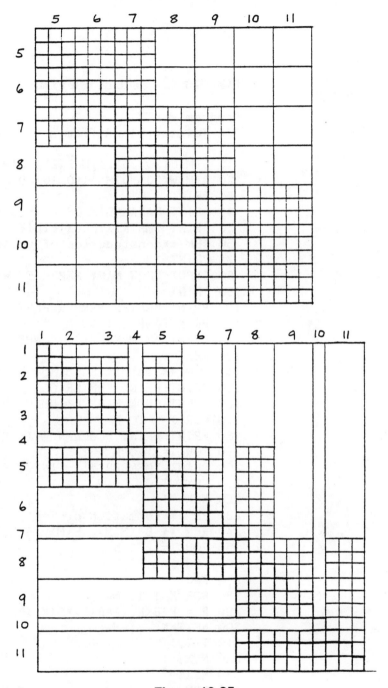

Figure 10.25

Solution — Fig. 10.25
(a) Bracketed set. (123 elements — upper diagram)
(b) Plain set. (139 elements — lower diagram)

BASIC listings

Program MFLEX1
Pin-jointed plane frames by flexibility method

```
 10 DIM B(20,20):DIM F(20):DIM T(20,20)
 20 DIM W(20):DIM A(10,10):DIM X(10)
 30 PRINT
 40 PRINT "ANALYSIS OF PIN JOINTED FRAMES"
 50 PRINT "MATRIX FLEXIBILITY METHOD"
 60 REM **construction of vector f **
 70 PRINT
 80 INPUT "HOW MANY MEMBERS (MAX 20)";M
 90 PRINT
100 PRINT "ENTER MEMBER FLEXIBILITIES L/EA"
110 PRINT:FOR I=1 TO M
120 PRINT "MEMBER ";I;: INPUT F(I)
130 NEXT I
140 REM ** matrix B col by col **
150 PRINT:PRINT "ENTER COLS OF MATRIX B"
160 INPUT "HOW MANY COLUMNS (MAX 20)";N
170 FOR J=1 TO N
180 PRINT:PRINT "COLUMN ";J:PRINT
190 FOR I=1 TO M
200 PRINT "MEMBER ";I;:INPUT B(I,J)
210 NEXT I
220 NEXT J
230 REM **formation of T=B'FB **
240 REM ** B'(N,M), F(M), B(M,N), T(N,N) **
250 FOR I=1 TO N
260 FOR J=1 TO N
270 P=0
280 FOR K=1 TO M
290 P = P+B(K,I)*F(K)*B(K,J)
300 NEXT K
310 T(I,J)=P
320 NEXT J
330 NEXT I
340 REM ** redundancies R; S=N-R **
350 REM ** input S loads into vector W **
360 PRINT
370 INPUT "HOW MANY REDUNDANCIES (MAX 10)";R
380 S=N-R
```

```
390 PRINT
400 PRINT "ENTER LOADS, INCLUDING ZEROS"
410 PRINT:FOR I=1 TO S
420 PRINT "ROW ";I;:INPUT W(I)
430 NEXT I
440 REM ** compute X (rhs of equations) **
450 PRINT
460 PRINT "ENTER LACK OF FIT AND SETTLEMENT"
470 PRINT:P=S
480 FOR I=1 TO R
490 P=P+1
500 PRINT "ROW ";P;:INPUT Z
510 FOR K=1 TO S
520 Z=Z-W(K)*T(P,K)
530 NEXT K
540 X(I)=Z
550 Q=S
560 FOR J=1 TO R
570 Q=Q+1
580 A(I,J)=T(P,Q)
590 NEXT J
600 NEXT I
610 GOSUB 1010
615 IF R=0 THEN GOTO 720
620 GOSUB 2520
630 PRINT
640 PRINT "REDUNDANT RESTRAINTS ARE:":PRINT
650 P=S
660 FOR I=1 TO R
670 PRINT X(I)
680 REM ** put redundancies in load vector **
690 P=P+1
700 W(P)=X(I)
710 NEXT I
715 PRINT:GOSUB 2000
720 PRINT:PRINT "MEMBER FORCES ARE:":PRINT
730 FOR I=1 TO M
740 P=0
750 FOR J=1 TO N
760 P=P+W(J)*B(I,J)
770 NEXT J
780 PRINT "MEMBER ";I;"     ";P
790 NEXT I
795 PRINT:GOSUB 2000
800 REM ** T*W for deflections **
810 PRINT
820 PRINT "DEFLECTIONS ARE:":PRINT
```

```
 830 FOR I=1 TO S
 840 P=0
 850 FOR J=1 TO N
 860 P=P+T(I,J)*W(J)
 870 NEXT J
 880 PRINT P
 890 NEXT I
 900 END
1000 REM ** print T, ie matrix F **
1010 FOR J=1 TO N
1020 PRINT: PRINT "MATRIX F COLUMN ";J
1030 PRINT:FOR I=1 TO N
1040 PRINT T(I,J)
1050 NEXT I:PRINT
1060 GOSUB 2000
1070 NEXT J:PRINT
1140 RETURN
2000 PRINT "PRESS SPACE BAR TO CONTINUE"
2010 IF INKEY$<>" " THEN GOTO 2010
2020 RETURN
2500 REM ** solve AY=X putting Y in X **
2510 REM ** A must be +ve definite **
2520 D=R-1
2530 FOR K=1 TO D
2540 U=K+1
2550 FOR I=U TO R
2560 P=A(I,K)/A(K,K)
2570 A(I,K)=0
2580 FOR J=U TO R
2590 A(I,J)=A(I,J)-P*A(K,J)
2600 NEXT J
2610 X(I)=X(I)-P*X(K)
2620 NEXT I
2630 NEXT K
2640 X(R)=X(R)/A(R,R)
2650 I=R
2660 FOR K=1 TO D
2670 I=I-1
2680 V=I+1
2690 Q=0
2700 FOR J=V TO R
2710 Q=Q+A(I,J)*X(J)
2720 NEXT J
2730 X(I)=(X(I)-Q)/A(I,I)
2740 NEXT K
2750 RETURN
```

Program MFLEX2
Rigidly jointed plane frames by flexibility method

This program is identical with MFLEX1 except for the lines marked ‡

```
 10 DIM BA(20,20):DIM BB(20,20):DIM F(20):DIM T(20,20)        ‡
 20 DIM W(20):DIM A(10,10):DIM X(10)
 30 PRINT
 40 PRINT "ANALYSIS OF RIGIDLY JOINTED FRAMES"                ‡
 50 PRINT "MATRIX FLEXIBILITY METHOD"
 60 REM **construction of vector f **
 70 PRINT
 80 INPUT "HOW MANY MEMBERS (MAX 20)";M
 90 PRINT
100 PRINT "ENTER MEMBER FLEXIBILITIES L/6EI"                  ‡
110 PRINT:FOR I=1 TO M
120 PRINT "MEMBER ";I;: INPUT F(I)
130 NEXT I
140 REM ** matrix B col by col **
150 PRINT:PRINT "ENTER COLS OF MATRIX B"
160 INPUT "HOW MANY COLUMNS (MAX 20)";N
170 FOR J=1 TO N
180 PRINT:PRINT "COLUMN ";J:PRINT
190 FOR I=1 TO M
200 PRINT "MEMBER ";I:PRINT                                  ‡
204 PRINT "END A";::INPUT BA(I,J)                            ‡
205 PRINT "END B";::INPUT BB(I,J):PRINT                      ‡
210 NEXT I
220 NEXT J
230 REM **formation of T=B'FB **
240 REM ** B'(N,M), F(M), B(M,N), T(N,N) **
250 FOR I=1 TO N
260 FOR J=1 TO N
270 P=0
280 FOR K=1 TO M
285 B1=BA(K,J):B2=BB(K,J)                                    ‡
290 P = P+F(K)*(BA(K,I)*(2*B1-B2)+BB(K,I)*(-B1+2*B2))        ‡
300 NEXT K
310 T(I,J)=P
320 NEXT J
330 NEXT I
340 REM ** redundancies R; S=N-R **
350 REM ** input S loads into vector W **
360 PRINT
370 INPUT "HOW MANY REDUNDANCIES (MAX 10)";R
380 S=N-R
390 PRINT
```

```
400 PRINT "ENTER LOADS, INCLUDING ZEROS"
410 PRINT:FOR I=1 TO S
420 PRINT "ROW ";I;:INPUT W(I)
430 NEXT I
440 REM ** compute X (rhs of equations) **
450 PRINT
460 PRINT "ENTER LACK OF FIT AND SETTLEMENT"
470 PRINT:P=S
480 FOR I=1 TO R
490 P=P+1
500 PRINT "ROW ";P;:INPUT Z
510 FOR K=1 TO S
520 Z=Z-W(K)*T(P,K)
530 NEXT K
540 X(I)=Z
550 Q=S
560 FOR J=1 TO R
570 Q=Q+1
580 A(I,J)=T(P,Q)
590 NEXT J
600 NEXT I
610 GOSUB 1010
615 IF R=0 THEN GOTO 720
620 GOSUB 2520
630 PRINT
640 PRINT "REDUNDANT RESTRAINTS ARE:":PRINT
650 P=S
660 FOR I=1 TO R
670 PRINT X(I)
680 REM ** put redundancies in load vector **
690 P=P+1
700 W(P)=X(I)
710 NEXT I
715 PRINT:GOSUB 2000
720 PRINT:PRINT "MEMBER FORCES ARE:":PRINT
730 FOR I=1 TO M
740 P=0:Q=0                                              ‡
750 FOR J=1 TO N
760 P=P+W(J)*BA(I,J)                                     ‡
765 Q=Q+W(J)*BB(I,J)                                     ‡
770 NEXT J
780 PRINT I;TAB(5);"A ";P;TAB(23);"B ";Q                 ‡
790 NEXT I
795 PRINT:GOSUB 2000
800 REM ** T*W for deflections **
810 PRINT
820 PRINT "DEFLECTIONS ARE:":PRINT
830 FOR I=1 TO S
```

```
 840 P=0
 850 FOR J=1 TO N
 860 P=P+T(I,J)*W(J)
 870 NEXT J
 880 PRINT P
 890 NEXT I
 900 END
1000 REM ** print T, ie matrix F **
1010 FOR J=1 TO N
1020 PRINT: PRINT "MATRIX F COLUMN ";J
1030 PRINT:FOR I=1 TO N
1040 PRINT T(I,J)
1050 NEXT I:PRINT
1060 GOSUB 2000
1070 NEXT J:PRINT
1140 RETURN
2000 PRINT "PRESS SPACE BAR TO CONTINUE"
2010 IF INKEY$<>" " THEN GOTO 2010
2020 RETURN
2500 REM ** solve AY=X putting Y in X **
2510 REM ** A must be +ve definite **
2520 D=R-1
2530 FOR K=1 TO D
2540 U=K+1
2550 FOR I=U TO R
2560 P=A(I,K)/A(K,K)
2570 A(I,K)=0
2580 FOR J=U TO R
2590 A(I,J)=A(I,J)-P*A(K,J)
2600 NEXT J
2610 X(I)=X(I)-P*X(K)
2620 NEXT I
2630 NEXT K
2640 X(R)=X(R)/A(R,R)
2650 I=R
2660 FOR K=1 TO D
2670 I=I-1
2680 V=I+1
2690 Q=0
2700 FOR J=V TO R
2710 Q=Q+A(I,J)*X(J)
2720 NEXT J
2730 X(I)=(X(I)-Q)/A(I,I)
2740 NEXT K
2750 RETURN
```

Program MSTIFF

Rigidly jointed and pin-jointed plane frames by stiffness method

```
10 PRINT "MATRIX STIFFNESS ANALYSIS": PRINT
20 OPTION BASE 1
30 DEFINT H-N,S-V,Z
40 DIM NX(20),NY(20),NT(20),LJ(20),JI(20)
50 DIM JJ(20),NC(20),S(60),X(20),Y(20)
60 DIM CA(30),CI(30),A(900),B(60),C9(60)
70 DIM J(10),WX(10),WY(10),WM(10)
80 PRINT "ENTER GENERAL DATA":PRINT
90 INPUT "YOUNG'S MODULUS"; YM
100 INPUT "HOW MANY MEMBERS";M
110 INPUT "HOW MANY JOINTS";N
120 INPUT "HOW MANY LOADED JOINTS";NW
130 PRINT:PRINT "ENTER JOINT DATA":PRINT
140 FOR K=1 TO N
150 PRINT "JOINT";K:INPUT ;"X";X(K)
160 INPUT " Y";Y(K):INPUT ;"D.O.F.  X";NX(K)
170 INPUT ;" Y";NY(K):INPUT" THETA";NT(K):
180 PRINT
190 REM ** Set all LJ to K **
200 LJ(K)=K
210 NEXT K
220 GOSUB 2170
230 PRINT:PRINT "ENTER MEMBER DATA"
240 FOR L=1 TO M
250 PRINT:PRINT "MEMBER";L
260 INPUT "C.S.A.";CA(L)
270 INPUT "S.M.A.";CI(L)
280 INPUT "JOINT i";JI(L)
290 INPUT "JOINT j";JJ(L)
300 NEXT L
310 GOSUB 2170
320 REM ** Form LJ array **
330 FOR L=1 TO M
340 I=JI(L):J=JJ(L)
350 IF I<J THEN K=I:I=J:J=K
360 IF LJ(I)>J THEN LJ(I)=J
370 NEXT L
380 REM ** Cycle joints **
390 REM ** Form NC and S arrays **
400 NC=0:S=0
410 FOR K=1 TO N
420 NC(K)=NC:NK=NC
430 NF=NX(K)+NY(K)+NT(K)
440 IF NF=0 THEN GOTO 1030
450 LJ=LJ(K):JD=NC-NC(LJ)
```

```
460 FOR I=1 TO NF
470 S=S+JD+I:S(NC+I)=S
480 NEXT I
490 NC=NC+NF
500 REM ** Set LD elements to zero **
510 D1=0:D2=0:D3=0:D4=0:D5=0:D6=0
520 REM ** Cycle members to find
530 REM ** connection with joint K **
540 FOR L=1 TO M
550 I=JI(L):J=JJ(L)
560 IF J=K THEN KK=I: GOTO 610
570 IF I<>K THEN GOTO 960
580 KK=J
590 REM ** Connection found K - KK **
600 REM ** Compute stiffness coeff's **
610 X=X(J)-X(I):Y=Y(J)-Y(I)
620 CL=SQR(X↑2+Y↑2)
630 C1=X/CL↑2:C2=Y/CL↑2
640 C3=CL*YM*CA(L):C4=2*YM*CI(L)/CL
650 A=C3*C1↑2+C4*6*C2↑2
660 B=C3*C2↑2+C4*6*C1↑2
670 C=C3*C1*C2-C4*6*C1*C2
680 D=-C4*3*C2:E=C4*3*C1
690 F=C4*2:G=C4
700 IF KK=I THEN D=-D:E=-E
710 REM ** Sum LD elements **
720 IF NX(K)=0 THEN GOTO 790
730 D1=D1+A
740 IF NY(K)=0 THEN GOTO 850
750 D2=D2+C:D3=D3+B
760 IF NT(K)=0 THEN GOTO 880
770 D4=D4+D:D5=D5+E:D6=D6+F
780 GOTO 880
790 IF NY(K)=0 THEN GOTO 840
800 D1=D1+B
810 IF NT(K)=0 THEN GOTO 880
820 D2=D2+E:D3=D3+F
830 GOTO 880
840 IF NT(K)=1 THEN D1=D1+F:GOTO 880
850 IF NT(K)=0 THEN GOTO 880
860 D2=D2+D:D3=D3+F
870 REM ** Feed rows of off-diag elements **
880 IF K<KK THEN GOTO 960
890 JD=NK-NC(KK):I=0
900 IF NX(K)=0 THEN GOTO 920
910 P=-A:Q=-C:R=D:GOSUB 1740
920 IF NY(K)=0 THEN GOTO 940
930 P=-C:Q=-B:R=E:GOSUB 1740
```

```
 940 IF NT(K)=0 THEN GOTO 960
 950 P=-D:Q=-E:R=G:GOSUB 1740
 960 NEXT L
 970 REM ** Feed sums of LD elements **
 980 S=S(NK+1):A(S)=D1
 990 IF NF<2 THEN GOTO 1030
1000 S=S(NK+2):A(S)=D3:A(S-1)=D2
1010 IF NF<3 THEN GOTO 1030
1020 S=S(NK+3):A(S)=D6:A(S-1)=D5:A(S-2)=D4
1030 NEXT K
1040 REM ** Construct load vector B **
1050 FOR I=1 TO NC:B(I)=0:NEXT I
1060 PRINT:PRINT "ENTER LOAD DATA":PRINT
1070 FOR I=1 TO NW
1080 INPUT "JOINT NO.";J(I):J=J(I)
1090 INPUT "X LOAD";WX(I)
1100 INPUT "Y LOAD";WY(I)
1110 INPUT "M LOAD";WM(I)
1120 PRINT
1130 NEXT I
1140 GOSUB 2170
1150 FOR I=1 TO NW
1160 J=J(I):K=NC(J)
1170 IF NX(J)=1 THEN K=K+1:B(K)=WX(I)
1180 IF NY(J)=1 THEN K=K+1:B(K)=WY(I)
1190 IF NT(J)=1 THEN K=K+1:B(K)=WM(I)
1200 NEXT I
1290 PRINT
1300 REM ** Solve equations **
1310 GOSUB 1930
1320 REM ** Output displacements **
1330 PRINT:PRINT "JOINT DISPLACEMENTS"
1340 L=0:S=3
1350 FOR K=1 TO N
1360 IF S>0 THEN GOTO 1380
1370 GOSUB 1880:S=3
1380 GOSUB 1800
1390 PRINT:PRINT "JOINT";K
1400 PRINT "X = ";WX
1410 PRINT "Y = ";WY
1420 PRINT "ROTATION = ";WM:S=S-1
1430 NEXT K:PRINT:PRINT
1440 GOSUB 1880
1450 PRINT:PRINT "MEMBER FORCES":PRINT
1460 REM ** Member joint displacements **
1470 S=3
1480 FOR I=1 TO M
1490 K=JI(I):L=NC(K):A=X(K):B=Y(K)
```

```
1500 GOSUB 1800
1510 C1=WX:C2=WY:C3=WM
1520 K=JJ(I):L=NC(K):C=X(K):D=Y(K)
1530 GOSUB 1800
1540 C1=WX-C1:C2=WY-C2:C6=WM
1550 REM ** Compute member forces **
1560 X=C-A:Y=D-B:CL=SQR(X↑2+Y↑2)
1570 X=X/CL:Y=Y/CL
1580 E=2*YM*CI(I)/CL
1590 P=YM*CA(I)/CL*(C1*X+C2*Y)
1600 C=3/CL*(Y*C1-X*C2)
1610 A=E*(C+2*C3+C6)
1620 B=E*(C+C3+2*C6)
1630 Q=-(A+B)/CL
1640 IF S>0 THEN GOTO 1660
1650 GOSUB 1880:S=3
1660 PRINT "MEMBER";I
1670 PRINT "P= ";P,"Q= ";Q
1680 PRINT "MA= ";A,"MB= ";B:PRINT
1690 S=S-1
1700 NEXT I
1720 END
1730 REM ** Sub to feed off-diag elements **
1740 I=I+1:S=S(NK+I)-JD-I
1750 IF NX(KK)=1 THEN S=S+1:A(S)=P
1760 IF NY(KK)=1 THEN S=S+1:A(S)=Q
1770 IF NT(KK)=1 THEN S=S+1:A(S)=R
1780 RETURN
1790 REM ** Sub to extract displacements **
1800 IF NX(K)=0 THEN WX=0: GOTO 1820
1810 L=L+1:WX=B(L)
1820 IF NY(K)=0 THEN WY=0:GOTO 1840
1830 L=L+1:WY=B(L)
1840 IF NT(K)=0 THEN WM=0:GOTO 1860
1850 L=L+1:WM=B(L)
1860 RETURN
1870 REM ** Sub to delay printing **
1880 PRINT
1890 PRINT "PRESS SPACE BAR TO CONTINUE"
1900 IF INKEY$="" THEN GOTO 1900
1910 RETURN
1920 REM ** Sub to solve equations **
1930 X9=A(1):V9=1
1940 FOR J9=1 TO NC
1950 U9=S(J9):IF J9=1 THEN GOTO 2070
1960 K9=U9-V9:I9=J9-K9:T9=0
1970 FOR H9=1 TO K9
1980 Z9=T9:V9=V9+1:I9=I9+1:T9=S(I9)
```

```
1990 M9=T9-H9+1:IF M9<=Z9 THEN M9=Z9+1
2000 N9=H9+M9-T9:X9=A(V9):GOTO 2020
2010 X9=X9-A(M9)*C9(N9):M9=M9+1:N9=N9+1
2020 IF M9<T9 THEN GOTO 2010
2030 C9(H9)=X9:A(V9)=A(T9)*X9
2040 IF H9=K9 THEN GOTO 2060
2050 B(J9)=B(J9)-B(I9)*X9
2060 NEXT H9
2070 X9=1/X9:A(U9)=X9:B(J9)=B(J9)*X9
2080 NEXT J9
2090 REM ** Back-substitution **
2100 FOR J9=NC TO 2 STEP -1
2110 U9=T9-1:K9=J9-1:T9=S(K9)
2120 IF T9=U9 THEN GOTO 2150
2130 X9=A(U9):B(K9)=B(K9)-X9*B(J9)
2140 U9=U9-1:K9=K9-1:GOTO 2120
2150 NEXT J9
2160 RETURN
2170 PRINT:PRINT "DIRECT MODE. CORRECT DATA"
2180 PRINT "ARRAYS IF NECESSARY":PRINT
2190 PRINT "ENTER CONT TO CONTINUE":PRINT:PRINT
2200 PRINT:STOP:RETURN
```

Index

73   23